Neural Networks for
Robotic Control

Neural Networks for Robotic Control
Theory and Applications

Edited by

A. M. S. Zalzala
and
A. S. Morris

Robotics Research Group
Department of Automatic Control and
Systems Engineering
University of Sheffield

ELLIS HORWOOD

LONDON   NEW YORK   TORONTO   SYDNEY   TOKYO   SINGAPORE
MADRID   MEXICO CITY   MUNICH

First published 1996 by
Ellis Horwood Limited
Campus 400, Maylands Avenue
Hemel Hempstead
Hertfordshire HP2 7EZ
A division of
Simon & Schuster International Group

Printed and bound in Great Britain by
Hartnolls Limited, Bodmin, Cornwall

Library of Congress Cataloging-in-Publication Data

Neural networks for robotic control : theory and applications / edited
   by A.M.S. Zalzala and A.S. Morris.
       p.   cm.
     Includes bibliographical references.
     ISBN 0–13–119892–0 (hbk)
       1. Robots–Control systems.   2. Neural networks (Computer science)
   I. Zalzala, A. M. S. (Ali M. S.)   II. Morris, Alan S., 1948–   .
   TJ211.35.N47   1996
   629.8′9263–dc20   95–14216
                                                                          CIP

British Library Cataloguing in Publication Data

A catalogue record for this book is available from
the British Library

ISBN 0–13–119892–0

---

   2   3   4   5   00   99   98   97

Contents

Preface

The motivation in producing this book has been to assemble a number of contributions from distinguished researchers in the field which collectively cover all the different topics within the general subject area of neural network applications in robot control. The book therefore provides a state-of-the-art overview of the application of connectionism in this field.

There has been strong research activity over the last few years and a large number of publications have emerged which describe the use of neural networks in robotics control. However, whilst the theoretical content of these publications is generally excellent, the practical value of many of them is severely limited by their failure to address the issues and difficulties involved in implementing the algorithms proposed in real systems. With this in mind, the contributions in this text have been carefully chosen to ensure that the practical implementation issues associated with the algorithms proposed are properly discussed.

The book starts with a broad look at the use of artificial neural networks in control applications in general. In the first chapter, Warwick explains the ability of neural networks to map non-linear functions flexibly and arbitrarily, particularly in multivariable systems. Neural networks are a powerful tool which can be trained either on- or off-line and exhibit, to an extent at least, fault tolerant characteristics. However, Warwick explains that extreme problems can be presented in respect of stability and approximation analysis, and careful examination of the structural properties of both the network and the controlled plant is required. A further cautionary note is the need to consider carefully whether a neural network is appropriate at all, and, if it is, whether it should be supported by other means of control. For example, a neural network is inappropriate for most simple linear control problems because it is essentially a non-linear mapping tool and would probably lead to poorer quality control than would a conventional linear controller. On the other hand, some problems can be so complex that a neural network controller is inadequate by itself and must be supported by high-level, decision-making features, i.e. rule-based or expert systems.

A major research area in the application of neural networks in robot control is addressed towards fixed-base robot manipulators, and the following four chapters in the text cover various aspects of articulated manipulator control. Morris and Khemaissia present a system identification scheme and an adaptive controller for robot arms, and compare the suitability and performance of various methods for updating the network. This is of particular importance considering the initialisation problems associated with back propagation and other schemes. Fukuda and colleagues propose a neural servo scheme which aims to control highly non-linear systems with uncertainties, using the self-organising capabilities of neural networks. Having proved the effectiveness of their algorithm for hybrid position/force control, the authors proceed to examine its applicability for stabbing control. Of significant importance are the authors' experimental results. In his chapter, Zalzala presents a novel approach for constructing adaptive neural structures based on knowledge of the model. This overcomes the difficulties about the decision on the network structure for a particular application and gives a further insight into the neural operation. The effectiveness of the scheme is illustrated by two case studies. The co-ordination of multiple systems is illustrated by Cui and Shin through a hierarchical controller which combines a neural-based predictor and a knowledge-based co-ordinator. In addition to the simulation of two 2-link arms manipulating a rigid body, the algorithm is tested in a case study which involves the collision avoidance for two revolute robots.

The following two chapters include contributions on mobile robots, which is an area of rapidly growing interest. Moreno and colleagues address integrating learning and control for autonomous mobile robots. Emphasising a hierarchical structure, their scheme starts with sensory information and situations at a lower level but incorporates experiences and lessons at a higher level. Experimentation on a laboratory-scale system shows the effectiveness of the method. In his work, Biewald describes a navigation system for the control of non-holonomic vehicle models using a back propagation through time algorithm. The neural controller is simulated for different cases of task execution and obstacle avoidance while using data feedback from the sensors.

Machine vision plays a major role in the development of the intelligent capabilities of robotic systems. Watanabe presents an ultrasonic 3D robot vision system based on a statistical framework and shows that probabilistic design methods for artificial neural networks are useful in practical applications. Kubota and Hashimoto present a method for detecting the position and orientation of the end-effector based on visual sensors, using a network which learns the non-linear relation between the image data and the control signals. In both contributions, experimental results are reported which show the effectiveness of the methods.

Evolutionary algorithms are very much in their infancy, and the contribution by de Garis presents a good introduction to the subject. In his chapter, the author discusses the possibility of building artificial brains containing a billion neurones in the near future. Concepts, implementation issues and justifications are included for what appears to be the ultimate goal in intelligent robotics.

The volume is concluded by the work of Chiel and colleagues on locomotion in a hexapod robot. This is an important area of current interest whose aim is to develop a means of robot location over rough terrain. Based on insect microbiology, the authors have developed a robust neural controller for a walking robot. In addition to simulations, an implementation on a six-legged machine is reported.

Ali Zalzala   Sheffield
Alan Morris   January 1995

1

An overview of neural networks in control applications

K. Warwick
Department of Cybernetics, University of Reading, Reading, UK

1.1 INTRODUCTION

Perhaps the most exciting and innovative technical development over the last ten years in the field of control has been the introduction of artificial neural network methods for modelling, identification and control [1,2]. As with any new topic it still suffers from a certain amount of hype: labelling it as something which is far better than all its predecessors, whilst as a topic it also suffers from attack by critics who insist that it is really nothing new and has nothing to offer. In this light, this chapter serves two purposes. Firstly, a scientific view is given as to the usefulness, or otherwise, of artificial neural networks in the field of control. Secondly, the chapter should serve as something of an introduction to the field for those wishing to get into the area.

The basic concept of artificial neural nets stems from the idea of modelling individual brain cells/neurons in a fairly simple way, and then connecting these models in a highly parallel fashion to offer a complex processing mechanism which exhibits learning in terms of its overall nonlinear characteristics. Because of the modelling operation, different model types have been derived, these ranging widely in their complexity and operation, e.g. some are analog whereas others are binary. This means that some of the models vary considerably when their operation is compared with that of actual brain cells.

Actual brain cells are not of one particular form, i.e. they are not all identical, whereas at present artificial neural networks tend to consist of one type of neuron. Further, the overall make up of a brain, in terms of connectivity and structure, is highly complex and not well understood, whereas artificial neural networks are generally well structured and simply coupled, thereby enabling the possibility of understanding their mode of operation.

A conclusion from numerous application studies, within a control systems environment, is that artificial neural networks can rarely be used in a black-box, stand-alone mode, in that they are not generally powerful or versatile enough to deal with more than two or three tasks, often being restricted in operation to one specific role. For an actual application, therefore, it should be remembered that, although a neural network may be employed as a key processing element, much interfacing, software jacketing and high level decision making features, e.g. expert or rule-based systems, must also be employed.

Important points to be considered are thus which tasks the neural network should be left to deal with and which should be handled by some alternative means. On the one hand, if the nature or complexity of the problem is greater than that with which the neural network can deal adequately, poor overall performance will result. This could mean either that limited information processing/interfacing/jacketing, is not sufficient, such that the neural network is effectively dealing with a different (wrong) problem, or the structural complexity of the neural network employed is not sufficient for the problem in hand. However, in some cases the neural network may be far too complex for the problem, i.e. the network is over-parameterized or functionally simply too powerful for the particular problem addressed. These points are looked at further in Section 5.

It is worth, at this stage, adding a note of caution insofar as attempting to employ a neural network solution to just about every control task possible. On the face of it there is nothing immediately wrong with this ploy, although the use of neural networks on problems to which they are not well suited is debatable, the big question being - which problems are neural network solutions well suited to? Hopefully this chapter will go a long way to answering this question, although some immediate comments can be made: One reason for the employment of a neural network architecture [3] is its inherent parallel processing nature, examples of this use being for mathematical operations such as least squares approximation or matrix algebra. However, neural networks generally provide a generalising, fuzzy solution when compared to a dedicated mathematical function processor; indeed if this were not the case it is difficult to see how the approach could be called a neural network. The speed advantage obtained by means of a parallel processing procedure can, of course, be delivered through the use of standard parallel processing architecture and not necessarily by means of a neural network related technique.

In terms of a control systems environment, the majority of practical controllers actually in use are both simple and linear, and are directed towards the control of a plant which is either reasonably linear or at least linearisable. The use of neural networks for such an environment is considered further in Section 1.5.2; however it is sufficient here to state that a neural network is usually a complex nonlinear mapping tool, and the use of such a device on relatively simple linear problems makes very little sense at all, being a case of over-kill [4]. The fact that neural networks have the capability of dealing with complex nonlinearities in a fairly general way is an exciting feature. By their nature, nonlinear systems are nonuniform and invariably require custom designed control schemes to deal with individual characteristics. No general theory deals comprehensively with the wide range of nonlinear systems encountered, so an approach, namely, neural networks, which can offer a fairly broad coverage is therefore extremely attractive.

At present, a negative aspect of the use of neural networks exists, however, in that they present extreme problems insofar as stability and approximation analysis is concerned [5], and, further, that it is often difficult to decide on the network structure, although some results are available [6]. Further use of neural networks for control is thus dependent on the fact that either the nonlinear system type is such that strict operational proofs are not necessary or that neural networks are the only or best technique available. Problems in the use of such networks are even greater in the field of adaptive computer control where a neural network can be part of an adapting feedback loop; however this is an area in which relatively few strong mathematical proofs exist; hence by considering a neural network approach there is very little to lose.

By viewing neural networks with control applications clearly in mind, a number of observations can be made [7], the first of these being perhaps the most significant:

(i)   Neural networks can flexibly and arbitrarily map nonlinear functions. Such networks are best suited for the control of nonlinear systems.

(ii)   Neural networks are particularly well suited to multivariable applications due to their ability to map interactions and cross-couplings readily whilst incorporating many inputs and outputs. However, in dealing with dynamic systems, a number of instants of one signal are presented at several inputs or realised at several outputs, i.e. sampled versions of individual signals must be dealt with. This can be accommodated by the use of delays within the network, though.

(iii) Networks can either be trained off-line and subsequently employed either on- or off-line, or they can be trained on-line as part of an adaptive control scheme or simply a real-time system identifier. Understanding of a network's mode of operation within an adaptive controller is, at the present time, extremely limited.

(iv)   Neural networks are inherently parallel processing devices which exhibit, to an extent at least, fault tolerant characteristics. Fast data processing is therefore achievable in a framework of graceful degradation.

It should also be noted that neural networks range widely in type, the selection of any particular network being dependent on the characteristics of the intended application, e.g. necessary accuracy required and overall problem complexity. Important features are as follows. How well can the network model the problem/system? How well can the network model be modified in terms of both structure and parameter values? Is off-line or on-line implementation required? This latter point is critical in that use of any practical controller invariably requires an analysis of the overall loop gain, and hence system stability, which is rather a difficult problem to tackle in the case of a neural network base.

In Sections 1.2 and 1.3, two types of neural network, namely multi-layer perceptrons (MLP) and radial basis functions (RBF) are looked at. Section 1.4 then follows with a discussion of learning algorithms, whilst in Section 1.5 it is considered how neural networks can be used to model/identify a system. Finally, in Section 1.6 possible controller realisations are looked into.

1.2 MULTI-LAYER PERCEPTRONS

The range of possible neural networks is broad; in fact almost any approximation technique which can be represented in the form of a network can be regarded as a neural network [8]. Much literature is available concerning the performance of different network types, a good example being [7]; however in this text the two most popular methods will be looked at in greater detail, namely (a) multi-layer perceptrons, and (b) radial basis functions. Apart from these two methods other common network modelling procedures can readily be found, especially CMAC (cerebellar model articulation controller) [9], cellular neural networks [10], and orthogonal basis functions. For the most part these other methods exhibit certain similarities to radial basis functions; in particular their inherently linear parameter base, and hence much of the discussion in Section 1.3 applies equally to these methods.

Multi-layer perceptrons [11] provide one arrangement for neural network implementation, by means of nonlinear relationships between, firstly, the network inputs to outputs and, secondly, the network parameters to outputs. Such a network consists of a number of neuron layers, n, linking its input vector, y, by means of the equation:

$$ y = \phi_n\left(W_n \phi_{n-1}\left(W_{n-1} \cdots \phi_1\left(W_1 u + b_1\right) + \ldots + b_{n-1}\right) + b_n\right) \tag{1.1} $$

in which W_i is the weight matrix associated with the ith layer, ϕ_i is a nonlinear operator associated with the ith layer and b_i indicate threshold or bias values associated with each node in the ith layer. In the majority of cases ϕ_i is a sigmoidal function for all n and is either an identical function on each layer or takes on a linear form in the output layer. Typically only three layers are used in total, an input layer, an output layer and one hidden layer, although the use of more hidden layers is frequently encountered. A three- layered network is shown in Figure 1.1, in which m inputs and m outputs are assumed to apply, although it is worth pointing out that it is by no means necessary for the number of inputs to equal the number of outputs.

If, for any neuron summation, the result is denoted by λ, then for a sigmoidal nonlinear function we have that either

$$ \phi(\lambda) = \frac{1 - \exp(-\lambda)}{1 + \exp(-\lambda)} \tag{1.2} $$

or

$$ \phi(\lambda) = \frac{1}{1 + \exp(-\lambda)} \tag{1.3} $$

both (1.2) and (1.3) being popular selections.

Other nonlinear functions frequently encountered are (a) threshold functions, (b) hyperbolic tangents, and (c) Gaussian relationships.

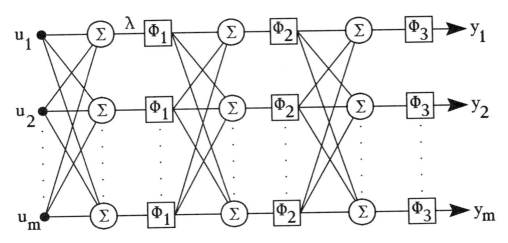

Figure 1.1 A three-layer multi-layer perceptron

It is known in reality that real neurons, located in different areas of the nervous system, have different modes of behaviour [12] ranging from Gaussian-like for visual needs to sigmoidal for occular motor needs. It is generally the case for artificial neural networks, however, that only one type of nonlinearity is employed for a particular network, this linking in closely with the fact that each network is only employed for one particular task.

Figure 1.1 shows a fully connected network in that all of the neuron outputs in one layer of the network are connected as inputs to the next layer. This is normal practice; however it is quite possible for part-connectivity to be realised by connecting a group of outputs to only specific inputs. By this means sub-models can be formed within the overall MLP, and these can be particularly useful where a specific system characteristic is to be dealt with or inputs/outputs from the system can be categorised into certain types.

Whatever the network connectivity, key questions in the use of MLPs are how many layers there should be and how many neurons there should be in each layer. Once these structural features of the network have been selected, it remains for the adjustable weights to be settled on such that the network is completely specified in terms of its functionality. It is usual for the weights to be adjusted in order to minimise the error between the actual output vector y and desired output vector y_d according to a previously defined criterion. This aspect is discussed in depth in Section 1.4 on learning algorithms. It is worth noting at this stage, however, that as the network is fully connected, by adjusting the weights in one layer this will have the knock-on effect of modifying all of the signals and hence the outputs in the following layers. It follows therefore that for a specific approximation or matching problem, all of the weights need to be adjusted to satisfy, in the best way possible, the input/output relationship.

The normal use of MLP networks involves, firstly, training the network on a set of data, possibly collected from a plant to be controlled. Once the network structure, i.e. number of layers and neurons, has been selected, then the network weights are adjusted so that the network's input/output relationship best approximates that obtained from the plant. Once this has been done the weights can be fixed and the network used as model of the plant under observation.   The neural network is in this case acting directly as a plant model, in the same way as a model obtained through system identification [13], with the structural decisions taken on the network directly reflecting those taken in a system identification exercise, the big difference being that the neural network is a much more complex modelling tool.

An important feature of multi-layer perceptron networks is that usually only a single nonlinear function, $\phi(.)$, is used, this acting on a set of linearly summed, weighted signals. Mathematically this simplifies any analysis carried out; however, the task is still not a simple one mainly due to the effect of a nonlinearity acting on a nonlinearity, as is the case with multiple layers.

1.3   RADIAL BASIS FUNCTION NETWORKS

Fairly recently, radial basis function (RBF) networks have been shown to function very efficiently, and to have the ability to model, in a fairly straightforward way, any arbitrary nonlinear system [14]. The output layer of such a network is merely a linear combination of the hidden layer signals, there being only one hidden layer. RBF networks thus offer considerable promise in terms of proving network stability and robustness, mainly because of the way in which the network can be readily defined by a set of equations.

With an input vector u, an element of the output vector y is given by:

$$y = \sum_{i=1}^{N} w_i R_i (u) + w_o \qquad (1.4)$$

in which w_i ($i = 1,2, ..., N$) are the network weights, w_o is a bias term and R_i are activation functions of the form

$$R_i (u) = \phi \left(\| u - c_i \| \right) \qquad (1.5)$$

where $c_i (i = 1, 2, ..., N)$ are a set of basis function centres and $\phi(.)$ is the radial basis function which can be chosen in one of a number of ways [3], popular examples being:

(a)   thin plate spline, where

$$\phi(r) = r^2 \log(r) \qquad (1.6)$$

or

(b)   Gaussian, where

$$\phi(r) = \exp[-r^2 / \sigma^2] \qquad (1.7)$$

in which σ is a scaling parameter and

$$r = \|u - c_i\| \tag{1.8}$$

i.e., r is the Euclidian distance from the input vector u to the centre location c_i.

Experience dictates that although the Gaussian function given in (1.7) appears to be the more intuitive, in that the neuron output will be larger the closer u is to the basis function centre, c_i, the thin plate spline does work very nicely in practice and can be recommended.

Further, where dynamic, sampled systems are being considered, in a discrete-time environment it is most likely that the input vector will take on a form such as

$$u = (y_{t-1}, y_{t-2}, \dots : u_{t-1}, u_{t-2}, \dots) \tag{1.9}$$

in which y_{t-1} and u_{t-1} represent the values of plant output and input respectively, one sample period previously.

In their most usual form, as shown in Figure 1.2, the centres, c_i, are fixed values and function learning, as for the MLP case, is carried out by adjustment of the weights, w_i. Because of the inherent linear output relationship properties, the weight adjustment can be carried out by means of a linear adjustment criterion, such as linear least squares, thereby realising a distinct advantage for RBF networks in general.

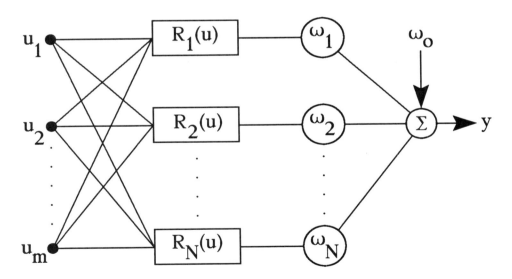

Figure 1.2 A radial basis function network

As was the case with MLP networks, a structural selection problem occurs with RBF networks, this time being dependent on how the centres are selected, both in terms of quantity and position. The choice of centres will, though, directly affect the quality of the function approximation obtained from the network, with a bottom line requirement that the number of centres employed must be such that they are sufficiently large in number to span the entire input domain [15]. However, the number of centres required tends to increase exponentially with regard to the input space dimension such that the RBF approach becomes unviable with a high dimensional space [16].

A direct trade-off exists between the number of RBF centres selected and the quality of function approximation obtained. For a low dimensional/complexity problem only a small number of centres are required: thus there is no point in over-parametrising the solution. However, where the problem is of high dimension, a reasonable solution can be provided by the selection of a fixed, fairly low number of centres, it being realised that the approximation provided by the RBF network is going to be, at best, reasonable. Once the number of centres has been selected and fixed, it then remains for the actual centre positions to be chosen, and this is perhaps better done as part of the network training exercise with the resultant values themselves being fixed. Allowing the centre values to be adjusted on-line, although opening up the possibility of network adaptability, realises extremely difficult problems, firstly with regard to proof of network stability, and secondly in terms of algorithmically controlling the nature and extent of centre adaptation.

In concluding this section on radial basis functions, the results of a simulation exercise are given in which the RBF network centres have been in turn chosen randomly, uniformly and by means of cluster analysis. Hopefully the case for selection by means of a clustering algorithm is thereby proven.

A simulation exercise was carried out based on one of the systems described in [11], i.e.

$$y_{t+1} = \frac{y_t y_{t-1}[y+2.5]}{1 + y_t^2 + y_{t-1}^2} + u_t \tag{1.10}$$

in which u_t is a random input in the range $[-1, 1]$.

The main choices for centre selection were taken as follows:

(a)   a random distribution of points chosen from within the convex hull of the input data set to be the RBF centres;

(b)   a uniform distribution of points chosen from within the convex hull of the input data set to be the RBF centres;

(c)   a clustering algorithm, as described in [5], was used to cluster the points in the input data set, with cluster centres being selected as RBF centres.

In each case, 100 data points were employed for network training with a further 50 data points subsequently used for testing purposes. Table 1.1 shows the mean squared error between the actual system output and the network output for each of the centre selection types described.   The number of clusters selected depends on p, a minimum specified distance between clusters value, where $0 < p < 1$.

Table 1.1 Accuracy of RBF network
($p = 0.001$ to $p = 0.5$ indicates clustered centres)

	Type of Centre	Number of Centres	Mean Squared Error
1.	Random	30	1.0692
2.	Uniform	30	0.2272
3.	$p = 0.5$	3	0.4454
4.	$p = 0.1$	27	0.0438
5.	$p = 0.05$	81	0.0170
6.	$p = 0.001$	100	0.0180

It can clearly be seen from Table 1.1 that when a clustering approach is used, the approximation accuracy improves as the number of centres is increased, until the stage where every data point is a cluster centre.

A further set of tests was conducted to investigate, for a set of centres chosen to be fixed at 28, which centring method was able to produce the best modelling accuracy over the 100 data points, and these results are given in Table 1.2.

Table 1.2 Centring accuracy over a given data set

Type of Centre	Mean Squared Error
Random	0.5136
Uniform	0.6936
Clustered	0.3658

It is apparent that a clustering approach appears to come out on top in both cases, although it should be pointed out that these results merely apply to one system simulation. It is clear to see, however, that RBF networks do provide a good potential solution for neural network based identification and control, although there is presently plenty of scope in finding more robust and reliable methods of basis function centre selection.

The two main neural network application modes have now been introduced.   In the section which follows, a look is taken at the learning algorithms employed for training network weights.

1.4   LEARNING ALGORITHMS

In the previous section, the two most popular schemes for neural network implementation, namely the multi-layer perceptron (MLP) and the radial basis function (RBF), were discussed, particularly in terms of their functionality as nonlinear mapping devices. In both cases, operation selection is necessary, firstly to determine the network's structure, and subsequently to determine the coefficient weightings. Structure and weighting selection can be carried out algorithmically; however the norm, mainly because of computational common sense, is for the structural selection to be made off-line, usually in a relatively *ad hoc* way, and then for the coefficient weightings to be adjusted in order to satisfy a cost functional algorithm. With weightings so adjusted, MLPs and RBFs can then both be used either as system identifier, modelling tool or as part of a nonlinear controller.

Coefficient weighting adjustment, termed learning in the neural network case, is based on the use of a relevant set of system input/output data, an attempt being made to adjust the network weights such that for the same input set the network provides, as near as possible, the same output set. Hence for a different input set, as long as the original training set is representative, the neural network model will behave as though it were the system itself.

Learning algorithms are generally based on the minimisation of an energy or error function, $E(W)$ where W is the vector of weights to be adjusted. In particular, gradient methods are widely applicable and have been shown to perform well in practice, although theoretical stability of such methods remains elusive. In this section gradient methods are firstly described in general and secondly in terms of implementation via a back propagation algorithm, the most popular learning tool for MLP weights.

1.4.1   Gradient descent algorithms

Consider the problem of minimising the scalar cost function: $E(W)$ by means of an iterative procedure generating a number of search points, $W(k)$, such that

$$W(k+1) = W(k) + \eta(k)d(k) \tag{1.11}$$

where an initial set of weightings $W(0)$ is made through prior knowledge, by a reasoned guess or even relatively randomly. The term $d(k)$ indicates the search direction, whereas $\eta(k)$ indicates the length of search step or the amount of learning to be carried out. From (1.11) it can be seen that with $\eta(k)$ very small, very little learning occurs as $W(k+1)$ is not changed much from $W(k)$, whereas when $\eta(k)$ is large, for a given $d(k)$, so $W(k+1)$ will be considerably different from $W(k)$.

For the steepest descent method the search direction is given by:

$$d(k) = -\nabla E(W(k)) \tag{1.12}$$

where ∇ indicates the derivative, or gradient of the error function.

For Newton's method, however, the search direction is found from

$$d(k) = -[\nabla^2 E(W(k))]^{-1} \nabla E(W(k)) \tag{1.13}$$

However, the evaluation of the inverse of $\nabla^2 E(W(k))$ may be an extremely complicated and computationally intensive task; hence an approximation of the matrix, defined as $H(k)$, is often used, and when used in the learning algorithm this becomes the quasi-Newton procedure with:

$$d(k) = -H(k)\nabla E(W(k)) \tag{1.14}$$

Finally, the conjugate gradient method employs a recursive form of search direction to give the equation:

$$d(k) = \alpha(k)d(k-1) - \nabla E(W(k)) \tag{1.15}$$

in which $\alpha(k)$ is a scalar parameter.

Of the methods described, the steepest descent (1.11, 1.12) is the simplest to implement; however convergence of the weighting coefficients may well be extremely slow. The Newton method (1.11, 1.13) provides a converse example in that although convergence is usually fairly rapid, when $W(0)$ is sufficiently close to the desired W, it is computationally expensive to implement, especially as a matrix inversion is involved. The quasi-Newton method is computationally a better proposition, although adequate memory storage for $H(k)$ must be provided, and fully conjugate gradient methods require more gradient evaluations but need less storage space when compared to quasi-Newton [17]. So it is very much a case of swings and roundabouts.

Finally, it is worth noting that the convergence rate of the algorithms described can be speeded up by applying a number of factors [18]. Firstly every element $w_i(k)$ of the vector $W(k)$ can have its own learning rate associated with it, which modifies the equations (1.11, 1.12) to:

$$w_i(k+1) = w_i(k) - \eta_i(k)\frac{\partial E(W(k))}{\partial w_i} \tag{1.16}$$

where $\eta_i(k)$ is a positive scalar value which should adapt using gradient information.

Further, if the expression $\partial E(W(k))/\partial w_i$ does not change sign over a number of steps, η_i can be increased as the minimum is still being approached from one direction, the converse being true when the sign changes on consecutive steps.

1.4.2   Back propagation learning

By far the most popular method employed for weight training in MLP neural networks is called back propagation. This procedure has been studied widely and is well reported on in the literature [19]. In the standard feedforward MLP network described in Section 1.2, back propagation solves the problem of missing information to the hidden layers, i.e. neither the input to nor the reference signals for the hidden layers are known, by taking the inputs to the hidden layers as being the inputs to the first layers propagated through the network. The reference signals for the hidden layers are then obtained by error back propagation through the network. This is realised by obtaining the partial derivative of the squared error with respect to the parameters.

It is worth pointing out that the back propagation algorithm has also been used for weight learning in feedback (recurrent) neural networks [20,21], these being networks in which the network structure incorporates feedback, whereby the output of every neuron is fed back, in weighted form, to the input of every neuron, i.e. the network is fully connected. The architecture of such a network is inherently dynamic and realises powerful capabilities due to its complexity. The discussion here on back propagation however is, for explanation purposes, limited to the feedforward, MLP network case.

Consider, as a starting point, a single neuron with output y_i; then

$$y_i = \frac{1 - \exp(-2\lambda_i x_i)}{1 + \exp(-2\lambda_i x_i)} = \phi(x_i) \tag{1.17}$$

in which

$$x_i = \sum_{i=1}^{n} w_{ij} u_j + w_0 \tag{1.18}$$

where w_0 is a bias term and λ_i are previously specified values denoting the type of sigmoidal function applied. If it is assumed that, at an instant in time, for an input u_i the output y_i should be equal to the desired output y_d, then the squared error of the output signal is given by:

$$E_i = \frac{1}{2}(y_d - y_i)^2 = \frac{1}{2}e_i^2 \tag{1.19}$$

and it is desired to minimise E_i by means of a suitable/best choice of the weighting coefficients w_{ij}. The steepest descent rule (1.11 , 1.12) gives a starting point as:

$$\frac{dw_{ij}}{dt} = -\eta \frac{\partial E_i}{\partial w_{ij}} \tag{1.20}$$

or, by taking x_i into account, this becomes:

$$\frac{dw_{ij}}{dt} = -\eta \frac{\partial E_i}{\partial x_i} \cdot \frac{\partial x_i}{\partial w_{ij}} \tag{1.21}$$

and hence

$$\frac{dw_{ij}}{dt} = \eta \delta_i u_i \tag{1.22}$$

in which

$$\delta_i = \frac{\partial E_i}{\partial x_i} = e_i \frac{d\phi}{dx_i} \tag{1.23}$$

On the basis of this algorithm, linking it in with (1.11, 1.12) we obtain that:

$$w_{ij}(k+1) = w_{ij}(k) + \eta_i(k)\delta_i(k)u_i(k) \tag{1.24}$$

in which

$$\delta_i(k) = e_i(k)\frac{d\phi}{dx_i}\bigg|_{x_i=x_i(k)} \tag{1.25}$$

To actually implement the learning equation (1.24), an accurate evaluation of the nonlinear sigmoid function ϕ is necessary, the shape of which depends directly on the selection of λ_i. In most cases this is not a distinct problem computationally, although it can present difficulties if an adaptive or real-time implementation of the algorithm is needed. One alternative is to employ a small perturbation signal Δx_i as an addition to the signal x_i, to serve as an estimation of the error function gradient. This gives a variation on (1.25) as

$$w_{ij}(k+1) = w_{ij}(k) - \eta_i(k)e_i(k)\left[\frac{\Delta e_i}{\Delta x_i}\right](k)u_i(k) \tag{1.26}$$

In this way, (1.25) or (1.26), the weights associated with one neuron can be adjusted in order to minimise the squared error (1.19). The approach can then be extended in order to adjust all of the weights in the MLP network. So, overall, a set of input/desired output data values is used to train the entire network, the input set also realising a corresponding set of network output values. The task is then to select network weights such that the error between the desired output signals and the actual network output signals is minimised in terms of the average overall learning points. The back propagation algorithm employs the steepest-descent method to arrive at a minimum of the mean squared error cost function. For one specific data pair, the error squared can be written as:

$$E_k = \frac{1}{2}\sum_{i=1}^{m}(y_{dk}-y_{ik})^2 = \frac{1}{2}\sum_{i=1}^{m}e_{ik}^2 \tag{1.27}$$

where m neurons are assumed to be present, and y_{ik} is the ith neuron's kth output value.

The global error is then found by minimising E_k over all the data set. If N data values are present, then we have:

$$E = \sum_{k=1}^{N}E_k \tag{1.28}$$

This error function can then be minimised in batch mode or recursively in an on-line manner. Looking at the latter of these cases for the neural network output layer only:

$$\Delta w_{ij} \; = \; -\eta \frac{\partial E_k}{\partial x_i} \frac{\partial x_i}{\partial w_{ij}} \tag{1.29}$$

and remembering (1.24):

$$w_{ij}(k+1) \; = \; w_{ij}(k) + \eta_i(k)\delta_i(k)u_i(k)$$

But for the output layer, the inputs $u_i(k)$ are in fact the outputs of the previous (hidden) layer, $\overline{u}_i(k)$, so:

$$w_{ij}(k+1) \; = \; w_{ij}(k) + \eta_i(k)\delta_i(k)\overline{u}_i(k) \tag{1.30}$$

in which

$$\delta_i(k) \; = \; (y_{dk} - y_{ik})\frac{\partial \phi}{\partial x_i}(k) \tag{1.31}$$

Now moving on to the hidden layer prior to the output layer, we have, as before, that

$$w_{ij}(k+1) = w_{ij}(k) + \eta_i(k)\overline{\delta}_i(k)\overline{u}_i(k) \tag{1.32}$$

where all weights, etc., refer to the hidden layer neurons. The main difference in fact lies in the way that $\overline{\delta}_i(k)$ is calculated, in that the global error signal taken in (1.31) is not the local error in (1.32). In fact, for this hidden layer:

$$\delta_i(k) \; = \; \frac{\partial \phi}{\partial x_i} \sum_{j=1}^{m} \delta_j(k)w_{ij}(k) \tag{1.33}$$

in which δ_i and w_{ij} refer to the output layer values whilst $\overline{\delta}_i(k)$ and ϕ refer to the hidden layer values. Input x_i is common. The error from the output layer therefore propagates through the hidden layer updating equation via (1.33), and this process is repeated until the input layer is reached, at which point the neuron weights for the input layer are given by:

$$w_{ij}(k+1) \; = \; w_{ij}(k) + \eta_i(k)\tilde{\delta}_i(k)u_i(k) \tag{1.34}$$

where $u_i(k)$ are now the overall network inputs and $\tilde{\delta}_i(k)$ is obtained from:

$$\tilde{\delta}_i(k) \; = \; \frac{\partial \phi}{\partial x_i}(k)\sum_{j=1}^{n} \overline{\delta}_j(k)w_{ij}(k) \tag{1.35}$$

in which n is the number of hidden layer neurons. Where more than one hidden layer exists, (1.35) relates to the first hidden layer in from the input, with n, $\overline{\delta}_j$ and w_{ij} all referring to that particular layer.

Overall then, by starting with the output layer, the error δ_i is calculated and is propagated back through the network layers until the input layer is reached. The only difference, therefore, between the weight updating calculations in the output layer and those in the hidden layers is that for the output layer the error depends on the difference between the desired and actual network output values, whereas for the hidden layers the local errors are calculated on the basis of the errors in the next (output) layer.

In a stepwise format, the back propagation algorithm can be implemented as follows:

1.   Select a suitable number of neuron layers, with a suitable number of neurons in each layer.

2.   Randomly select initial weightings w_{ij} throughout the network.

3.   Present a set of experimental input/desired output data to the network and measure the network's output to that data.

4.   Calculate $\overline{\delta_i}$ for the output layer and consequently back propagate to give this variable for all layers, concluding with the input layer, i.e $\overline{\delta_i}$, $\widetilde{\delta_i}$, ...

5.   Adjust the network weights to minimise E_k for each and every layer.

The network is now fully trained on the data presented and can be employed with any further data, although it may be desirable to present the data again cyclically until the overall error falls below a previously defined minimum value, i.e. until the weights converge. An important feature then emerges in that the MLP network has the ability to generalise when it is presented with new data not previously dealt with.

Quite a few modifications have been applied to the standard back propagation algorithm, one of them being to smooth the recursive weight changes by means of the updating equations (1.36, 1.37).

$$w_{ij}(k+1) \;=\; \beta(k) \,+\, w_{ij}(k) \qquad\qquad (1.36)$$

in which:

$$\beta(k) \;=\; \alpha\beta(k-1) \,+\, \eta\delta_i u_i \qquad\qquad (1.37)$$

where δ_i, u_i refer to values corresponding to the layer in which the neurons are being updated. The α term in (1.37) is called a momentum term and has a typical value of $\alpha = 0.9$. This in fact ties in very closely with the conjugate gradient method described earlier, accepting that α is now operator selected as opposed to being calculated explicitly. The effect of α is really to increase the learning rate if the previous weight change was large.

Other methods used to modify the speed of convergence of the back propagation algorithm involve either (a) applying a more sophisticated procedure, although this has computational drawbacks, (b) pre-processing the data or (c) obtaining a good initial estimate of the network weight values. Various heuristic methods also exist for improving the convergence speed [22] although the trade-off between convergence speed and computational effort becomes only too apparent; a number of such schemes are, however, well reviewed in [3].

1.4.3 Learning for RBF networks

The backpropagation algorithm just described is suitable for MLP network weight training. Training of weights in an RBF network is a much simpler task, being based on standard least squares algorithms [13], and only three steps are necessary:

1.   Select the number of RBF centres.
2.   Choose the centre values.
3.   Adjust the network weights.

Points 1 and 2 were discussed in more detail in Section 1.3 and will not be covered further here, although suffice to say that selection of the number and value of RBF centres is presently a fairly open problem and yet network performance depends critically on the selection made. Once the centres have been specified, however, adjustment of the network weights can be carried out as follows:

Assuming there to be N RBF centres, of value c_i, each data point is associated with its nearest centre. It is the case, therefore, when a clustering technique has been used to obtain the centre values, that each centre will have a different number of data points associated with it. The distance scaling parameters σ_i are then obtained heuristically, one way being by choosing a value p to indicate a number of neighbouring centres to any one particular centre. It follows then that:

$$\sigma_i = \frac{1}{p} \sum_{j=1}^{p} \sqrt{\|c_j - c_i\|^2} \qquad (1.38)$$

in which c_j is one of the p nearest neighbours to the centre c_i.

The weights of the RBF network, w_i, can then be found from the linear regression:

$$\hat{y}(k) = w_0 + \sum_{i=1}^{N} w_i R_i(u) \qquad (1.39)$$

where $R_i(u) = \phi(\|u - c_i\|)$ and n is as defined in Section 1.3.

By applying the linear least squares solution, this gives:

$$w_i(k+1) = w_i(k) + \mu_i e(k) R_i(u(k)) \qquad (1.40)$$

in which μ_i is positive definite and $e(k) = y(k) - \hat{y}(k)$, where $y(k)$ is the measured data output for the input vector $u(k)$.

A big adantage of RBF networks over MLP networks lies in the fact that weight training/learning times are about three times faster than for the back propagation algorithm, as far as networks of similar performance are concerned.

1.5 SYSTEM IDENTIFICATION

Neural networks are nonlinear function mapping devices. In the section which follows, a look is taken into how they can be employed within a control systems environment. One area of immediate interest however, which is directly addressed here, is their use as a plant modelling device, i.e. a system identifier.

A standard neural network identifier is shown in Figure 1.3, where network training is based on the use of plant input/output data, the overall aim being to obtain a neural network which performs, to all intents and purposes, in an identical way to the plant itself, the network acting as a nonlinear functional approximator.

It is well known that a continuous function can be represented, arbitrarily well,

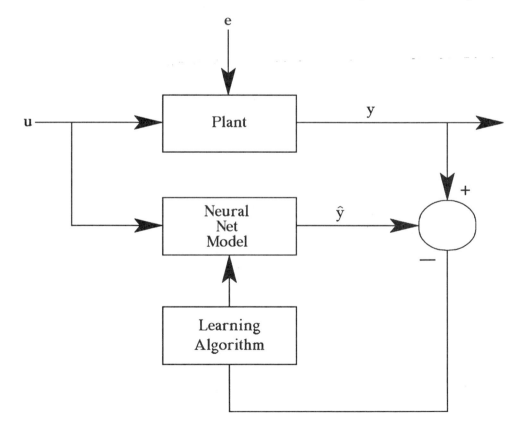

Figure 1.3 Neural net for system identification

by a series of polynomials. In a similar way, it has been shown [23] that an MLP network can approximate, arbitrarily well, a continuous function, and moreover only one hidden layer is required in order to do so, given that each unit in the hidden layer has a continuous, sigmoidal nonlinearity.

From a mathematical point of view, there is no immediately obvious advantage in the use of neural networks for modelling, as opposed to polynomials, although for each individual case it may be that one approach results in a simpler technique and therefore wins on that occasion. Implementation issues may also be taken into account and here it is likely that the neural network approach will come out on top, due to its inherent parallel nature.

What would be useful in the neural network case is an idea about how many hidden unit layers should be selected and how many neurons there should be in each layer, in order to obtain a desired accuracy of approximation. Purely applying theoretical results does not really achieve a solution to this problem [7] and perhaps at this stage a mixture of theory with heuristics [24] provides the most sensible solution.

A key property of a neural network in modelling a particular plant is perhaps not one of arbitrary approximation, but rather one of best approximation [8,14]. In this way a network has the best approximation property when it realises an approximation which is the minimum distance from the function to be modelled. It was in fact shown [14] that RBF networks do exhibit the best approximation property whereas MLP networks do not. Further, RBF networks can always be constructed by using only one hidden layer and their weights trained by using a linear technique which provides a guaranteed global solution [25]. It must be remembered, however, that choosing RBF network centres in a reasonable way provides something of a headache and realistically moves a practical RBF network away from the best solution unless a fortunate choice of number and value of centres has been made. So although, in theory, RBF networks have the ability to provide the best approximation, one must consider the practicalities involved and the cost incurred in using the method.

One problem to be considered is that most plants for which a system identification exercise is required will be dynamic in nature, and hence if they are to be modelled accurately the modelling tool itself must contain dynamics. One possibility is the construction of recurrent networks involving internal network feedback. Another possibility is to introduce dynamic behaviour into the neurons themselves [26]. The most usual way, however, is to force feed the network input with signals corresponding to previously periodically sampled plant input and output signals.

1.5.1 Nonlinear ARMA modelling

Under certain conditions any nonlinear function can be modelled by a Wiener or Volterra series. As a starting point consider the moving average (MA) representation:

$$y(k+1) = \sum_{i=0}^{\infty} a_i u(k-i) \qquad (1.41)$$

in which the a_i are scalar values indicating the system characteristics leading to the impulse response of the system due to a unit step application at time zero. Both the Wiener and

Volterra series generalise the linear representation (1.41) into nonlinear form such that the result is a linear time-invariant dynamical system operated on by a nonlinear transformation. The Volterra series, shown in (1.42), includes linear combinations of past input values taken multiplicatively with respect to time:

$$y(k+1) = a_0 + \sum_{i=0}^{\infty} a_i u(k-i) + \sum_{i=0}^{\infty} \sum_{j=0}^{\infty} a_{ij} u(k-i) u(k-j) + ... \qquad (1.42)$$

in which the a_0, a_i, a_{ij}, etc., are the unknown coefficients.

Neural networks including at least one hidden layer can arbitrarily approximate any nonlinear mapping, so the neural network shown in Figure 1.4 can represent any nonlinear function that can also be accurately modelled by a Volterra series, on condition that all of the past input values are available as network inputs.

Astute readers will spot that the neural network shown in Figure 1.4 does not contain an infinite number of input signals. In fact, as is normally the case with system identification, only a finite number of terms are used, such that a reasonable level of accuracy is achieved. Looked at another way, if n past input signals are applied to the network, then it is assumed that the input signal applied immediately prior to n sample periods ago has negligible effect on the present output signal.

Autoregressive (AR) models are also, in some cases, a useful way of representing the behaviour of a system, and in this case the plant output is linearly related to previous plant outputs. By restricting the model to n previous signals, the output is given by:

$$y(k+1) = \sum_{i=0}^{n-1} a_i y(k-i) \qquad (1.43)$$

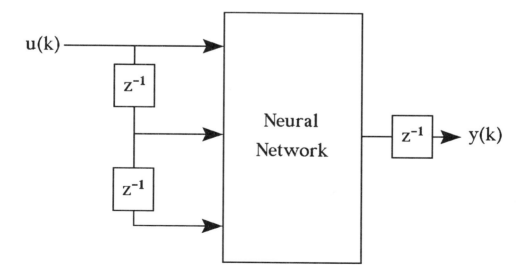

Figure 1.4 Neural network for MA modelling

and this form of model is useful for a time series description, involving a finite number of parameters, in which no input signal is directly applied. In nonlinear format, i.e. for nonlinear systems, the actual system output can be obtained simply by applying the appropriate nonlinear function to the summation obtained in (1.43).

Given a particular time series signal for the output $y(i)$, the neural network connection shown in Figure 1.5 can be used to obtain the functional output in which the neural network is employed to approximate any nonlinear effects, with network inputs simply using the a_i values as weights applied to the input signals. The plant itself can, in Figure 1.5, take on the form of a linear AR relationship, given that (a) $\hat{y}(k+1)$ represents $y(k+1)$ through a nonlinear function, (b) the plant can be adequately modelled by an nth order AR model and (c) no noise is corrupting the system and/or model.

For both MA and AR models, although an MLP is quite a feasible solution, a radial basis function network is only really sensible when the value of n, i.e. the number of past input/output values required, is relatively low.

Finally, by combining AR and MA models, it is known that any linear dynamical system of order n, looked at in discrete-time, can be represented by the difference equation:

$$y(k+1) = \sum_{i=0}^{n-1} a_i y(k-i) + \sum_{i=0}^{n-1} b_i u(k-i) \qquad (1.44)$$

which is the usual autoregressive moving average (ARMA) model, this subsuming AR and MA models as mere special cases.

By passing the result of (1.44) through a nonlinear function, and assuming that the real (plant) output is corrupted by noise, this defines the NARMAX plant model [27], and this representation has been used at length as a basis for neural network

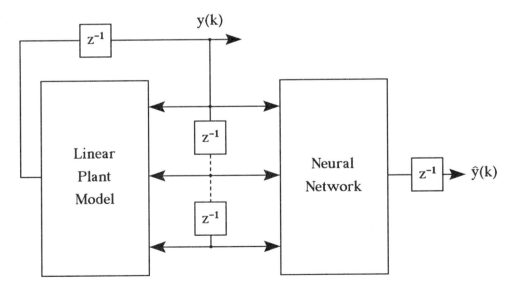

Figure 1.5 Neural network for AR modelling

approximation techniques, the aim of the network being to model the underlying nonlinear dynamics.

As far as system identification is concerned, neural networks have a number of advantages in that they have very good approximation properties, are simple and are straightforward to use, with parallel processing and fault tolerant characteristics. They are, however, merely a tool for approximating nonlinear functions. In order to apply a neural network for system identification, previously known system structure and parameter estimation results are usually ignored and this could well be at the operator's peril. As an example, consider identifying a first order heating system with a multi-layer perceptron. The model may well work extremely well; however, the essential simplicity of the original plant tends to be lost in the network weights and layers. It is therefore perhaps sensible to employ a neural network, not as a stand-alone for system identification, but rather as one part of a toolbox to be applied.

1.5.2   Linear neural networks

Whilst the potential is evident for multi-layer perceptron networks and, even more so, radial basis function networks in the identification and control of nonlinear systems, at the other end of the scale the potential of linear neural networks (LNN) has also been investigated [28]. In this section a brief look is taken at linear neural networks and a critical view is given.

Having recognised that neural networks are useful in tackling nonlinear problems, it should be remembered that the vast majority of practical controllers in use are both linear and simple, being based on the control of a plant which is either linear or linearisable. Hence, a number of authors have, in fact, suggested the use of neural networks for such as PID control [28].

A neural network is usually a complex nonlinear mapping tool and use of such a tool, in its general form, on relatively simple linear problems would make no sense at all, being very much a case of over-kill. By considering a linear form of neural network however, i.e. a much simplified approach, it could possibly be worth considering the employment of linear neural networks for linear control. One common approach for the use of such networks in control [29] is in terms of a modelling technique in order to obtain an approximation to the plant's inverse dynamics.   The overall controller then includes both a linear inverse dynamics model and a linear feedforward controller.

A linear neuron is essentially a summing junction such that, where an ARMA system model is being investigated, neuron inputs are in terms of delayed input and output values, as in (1.44), and model error (y) terms. The overall linear neural network control algorithm then consists of the following steps:

(a)   Sample the plant output, y.
(b)   Calculate $(y - \hat{y})$, where $\hat{y}$ is a model estimate of the output signal based on a linear neural network model of the plant.
(c)   Obtain a new control input where this is found from a linear neural network control equation.
(d)   Apply the control input.
(e)   Wait for the clock pulse then return to (a).

Unfortunately, when comparing the actual implementation of a LNN with an equivalent digital controller, [4], it is apparent that many more parameters are used within the overall scheme; in fact a factor of $2(n+2)^2$ parameters is used instead of n in the conventional case. This results in poor controller parameter network weight convergence for the LNN and an undesirable increase in computational effort. Overall, therefore, the use of linear neural networks is not recommended.

1.6 CONTROLLER REALISATIONS

It is worth concluding this article by looking at some basic implementation procedures for neural networks, within a control systems framework. As discussed in the previous section, in the first instance a system identification exercise can be carried out by means of an arrangement, as shown in Figure 1.3, in which e indicates the disturbance affecting the plant being identified and $(y-\hat{y})$ depicts the modelling error. Essentially, the model weights are adjusted in order to minimise the modelling errors.

    Neural networks can also be arranged in a number of ways within a feedback control scheme. One approach which is more open to stability and robustness analysis is internal model control (IMC), see e.g. [30], and in this method a system model is operated in parallel with the plant under control, as shown in Figure 1.6. The error between the plant and model outputs is then used as the feedback term. In this case the neural network model could have been obtained off-line through a plant identification exercise or from ideal model performance requirements.

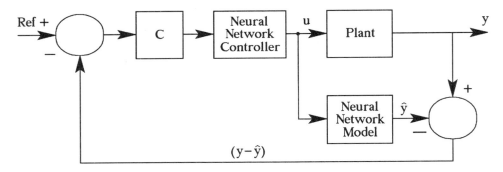

Figure 1.6 Internal model control
    The IMC feedback signal obtained is passed through a controller arrangement which is directly related to an inverse model of the plant, along with a sub-system C which is included for enhancing robustness and tracking capabilities. For a neural network realisation, two networks are thus used to apply the overall IMC technique, one modelling the forward plant characteristics and the other modelling the plant inverse characteristics. However, it must be pointed out that, principally because of the inverse model within the

loop, actual implementation is only directed towards open-loop stable systems; thus process control systems are particularly well suited.

Other control methods, e.g. optimal and predictive control, can be implemented through a neural network structure; the IMC case has merely served as an example. Neural networks can also be employed in a related way, such as for supervisory/decision making control at a higher level or for fault diagnosis or inferencing. An important overall point is whether the network is part of an automatic feedback loop or whether a human enters the loop in some way. If a fault diagnosis network is merely reporting on poor operating conditions or the nature and position of the fault, this can be treated rather like open-loop information for which a human operator can decide whether to believe or trust the information or not. Network operability, in terms of robustness, convergence, etc., is then not of such importance as it is within a closed-loop arrangement.

1.7 CONCLUSIONS

In this chapter, a broad look has been taken at the use of artificial neural networks specifically for modelling and control of nonlinear systems. It has been shown how such networks are useful as nonlinear mapping devices and links between the fields of neural networks and control were indicated. Radial basis function networks appear to be particulary promising, although much work still needs to be done in evaluating schemes for basis function centre selection. This can be seen as an existing gap in the field preventing multiple applications; however it can also be regarded as an exciting gap for researchers to fill, in terms of sensible solutions. It was also indicated how linear networks are essentially not particularly beneficial, in that linear control is well founded on relatively simple mathematical relationships such that the complexity of neural networks has nothing additional to offer.

A critical requirement for the implementation of neural networks within feedback control loops, particularly for adaptive and/or on-line application, is the need for detailed stability analysis results, in particular relating to the structural properties of the network and the plant under control. It is felt, however, that a more rigorous understanding of RBF network operation is first required before such stability results can be obtained, particularly in terms of centre selection and clustering algorithms.

Finally, for those who are new to the field, of the references which follow [2] is perhaps best as an introductory text in the area of neural networks for control, whereas [7] gives a more technically in-depth look at some different algorithms, although a better view of RBF networks is perhaps given in [3].

ACKNOWLEDGEMENTS

The author would like to thank various bodies for their research support; these include the UK Engineering and Physical Sciences Research Council, the UK Department of Trade and Industry, Molins plc, National Grid, Smith Kline Beechams, BP, Sun Microsystems,

National Power, UK Ministry of Defence, Amerace, British Telecom and several more in confidence.

REFERENCES

[1] W.T. Miller, R.S. Sutton, and P.J. Werbos (eds.), *Neural networks for control*, MIT Press, Cambridge, MA (1990).
[2] K. Warwick, G.W. Irwin, and K.J. Hunt (eds.), *Neural networks for control and systems*, Peter Peregrinus Ltd (1992).
[3] A. Cichocki and R. Unbehauen, *Neural networks for optimization and signal processing*, John Wiley and Sons (1993).
[4] K. Warwick, Neural networks for control: counter arguments, *Proc. IEE Int. Conference Control 94*, Warwick University, 95-99 (1994).
[5] J.D. Mason, R. Craddock, P. Parks, J.C. Mason and K. Warwick, Towards a stability approximation theory for neuro-controllers. *Proc. IEE Int. Conference Control 94*, Warwick University, 100-103 (1994).
[6] D.C. Psichogios and L.H. Ungar, SVD-NET: An algorithm that automatically selects network structure. *IEEE Trans. on Neural Networks*, **5**, 514-515 (1994).
[7] K.J. Hunt, D. Sbarbaro, R. Zbikowski and P.J. Gawthrop, Neural networks for control systems - a survey, *Automatica*, **28**, 1083-1112 (1992).
[8] T. Poggio and F. Girosi, Networks for approximation and learning, *Proc. IEEE*, **78**, 1481-1497 (1990).
[9] J.S. Albus, Data storage in the cerebellar model articulation controller (CMAC), *Trans. of the ASME Journal of Dynamic Systems, Measurement and Control*, **97**, 228-233 (1975).
[10] L.O. Chua and L. Yang, Cellular neural networks, *IEEE Trans. on Circuits and Systems Theory*, 1257-1290 (1988).
[11] K.S. Narendra and K. Parthasarathy, Identification and control of dynamical systems using neural networks. *IEEE Trans. on Neural Networks*, **1**, 4-27 (1990).
[12] D.H. Ballard, Cortical connections and parallel processing: structure and function. In *Vision, brain and cooperative computation*, M. Arbib and Hamson (eds.), 563-621, MIT Press, Cambridge, MA (1988).
[13] T. Soderstrom and P. Stoica, *System identification*, Prentice Hall (1989).
[14] F. Girosi and T. Poggio, Neural networks and the best approximation property, *Biol. Cybernetics*, **63**, 169-176 (1990).
[15] S. Chen, S.A. Billings, C.F. Cowan and P.M. Grant. Practical identification of NARMAX models using radial basis functions, *Int. Journal of Control*, **52**, 1327-1350 (1990).
[16] K.S. Narendra, Adaptive control of dynamical systems using neural networks, in *Handbook of intelligent control*, Van Nostrand (1994).
[17] D. Shanno, Recent advances in numerical techniques for large-scale optimisation, in *Neural networks for robotics and control*, W.T. Miller (ed.), MIT Press, Cambridge, MA, pp. 171-178 (1990).

[18] R.A. Jacobs, Increased rates of convergence through learning rate adaptation, *Neural Networks*, **1**, 295-307 (1988).

[19] D.E. Rumelhart and J.L. McClelland (eds.), *Parallel distributed processing: explorations in microstructure of cognition, 1: Foundations*, MIT Press, Cambridge, MA (1986).

[20] F.J. Pineda, Recurrent back propagation and dynamical approach to adaptive neural computation, *Neural computation*, **1**, 162-172 (1989).

[21] K.S. Narendra and K. Parthasarathy, Gradient methods for the optimization of dynamical systems containing neural networks, *IEEE Trans. on Neural Networks*, **2**, 252-262 (1991).

[22] S. Shah, F. Palmieri and M. Datum, Optimal filtering algorithms for fast learning in feedforward neural networks, *Neural Networks*, **5**, 779-787 (1992).

[23] G. Cybenko, Approximation by superpositions of a sigmoidal function, *Math. Control Signal Systems*, **2**, 304-314 (1989).

[24] R.P. Lippmann, An introduction to computing with neural networks, *IEEE ASSP Magazine*, **4**, 4-22 (1987).

[25] D.S. Broomhead and D. Lowe, Multivariable functional interpolation and adaptive networks, *Complex Systems*, **2**, 321-355 (1988).

[26] M.J. Willis, C. Di Massimo, G.A. Montague, M.T. Tham and A.J. Morris, On artificial neural networks in process engineering, *Proc. IEE*, Part D, **138**, 256-266 (1991).

[27] S. Billings and S. Chen, Neural networks and system identification, in *Neural Networks for Control and Systems*, K. Warwick, G.W. Irwin and K.J. Hunt (eds.), Peter Peregrinus Ltd, 181-205 (1992).

[28] K. Watanabe, T. Fukuda and S.G. Tzafesastas, An adaptive control for CARMA systems using linear neural networks, *Int. Journal of Control*, **56**, 483-497 (1992).

[29] W. Li and J.E. Slotine. Neural networks control of unknown nonlinear systems, *Proc.1989 American Control Conference*, Pittsburgh, 1136-1141 (1989).

[30] C.E. Garcia and M. Morari. Internal model control-1: A unifying view and some new results, *Ind. Eng. Chem. Process Des. Dev.*, **21**, 308-323 (1982).

2

Artificial neural network based intelligent robot dynamic control

A. S. Morris and S. Khemaissia
University of Sheffield, Department of Automatic Control and Systems Engineering, Robotics Research Group, Sheffield, United Kingdom

2.1 INTRODUCTION

Robotic manipulators have become increasingly important in the field of flexible automation. High-speed and high-precision trajectory tracking are indispensable capabilities for versatile applications of manipulators. Even in well-structured industrial applications, manipulators are subject to structured and/or unstructured uncertainties. Structured uncertainty is characterized by having a correct dynamical model but with parameter uncertainty due to imprecision of the manipulator link properties, unknown loads, inaccuracies in the torque constants of the actuators, and so on. Unstructured uncertainty is characterized by unmodelled dynamics. Unmodelled dynamics result from the presence of high-frequency modes of the manipulator, neglected time-delays, nonlinear friction, and so on.

When the dynamic model of the system under study is not known *a priori* (or is not available), the system has to be identified and then a certain control law is constructed based on this estimated model. This forms the basic idea behind adaptive- and learning-control strategies. Over the past few years, there has been considerable progress in understanding the functionalities which can be attained with artificial neural network computing, and the procedures and architectures available for realizing such functionalities. Advances have also been made in understanding how such functionalities might be combined and configured to support the performance of various application tasks, one of which is the control of systems.

In this chapter, we discuss the use of neural network computing for the intelligent dynamic control of robot systems. By this we mean that we are concerned with neural networks which can learn to control the system in accordance with a guiding intent, and also can learn to formulate that control strategy or intent.

Accordingly, the discussion will proceed in the following manner.

(a)   We will deal with the mechanistic aspects of the situation; namely, given a system and the means for observing and controlling it, how do we learn to control it so that it performs in the desired manner?

(b)   We will ask how we know what to specify as the desired performance, in detail, in a mode which can be utilized in control.

Our discussion is concerned with how such tasks might be carried out using neural network computing in an autonomous manner without reliance on arbitrary amounts of additional heuristics of human origin.

The novel system identification scheme and adaptive-control algorithm described is based on the computational properties of artificial neural network (ANN) models. It is applicable to a class of nonlinear systems, with the particular application in this case being the direct dynamic control of robot manipulators. The performance of various algorithms for updating the network parameters is compared. The techniques reviewed include backpropagation, recursive prediction error, recursive least squares and an extended Kalman filter algorithm.

A convenient starting point for the development of an ANN controller is the computed-torque algorithm. The computed-torque method is a well-established robot control technique, which takes account of the dynamic coupling between the robot links. Its main disadvantage is the assumption of an exactly known dynamic model, which is not realizable in practice. However, the basic algorithm remains important, as it forms the basis of various adaptive and neural controllers which have been developed to overcome the requirement for an exactly known dynamic model.

2.2 COMPUTED-TORQUE METHOD

For this method, the robot manipulator is modelled as a set of n rigid bodies connected in series with one end fixed to the ground and the other end free. The bodies are connected via either revolute or prismatic joints and a torque actuator acts at each joint. The dynamic equation of the manipulator is given by:

$$T = M(q)\ddot{q} + V(q,\dot{q}) + F(\dot{q}) + G(q) + T_d \tag{2.1}$$

where

T: $(n \times 1)$ vector of joint torques supplied by the actuators

$M(q)$: $(n \times n)$ manipulator inertia matrix

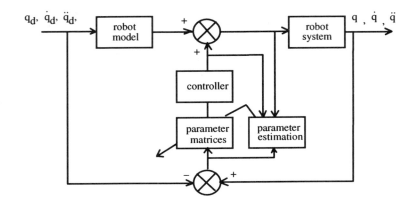

Figure 2.1 Architecture for computed-torque controller

$V(q,\dot{q})$: $(n\times 1)$ vector representing centrifugal and Coriolis effects

$F(\dot{q})$: $(n\times 1)$ vector representing friction forces

$G(q)$: $(n\times 1)$ vector representing gravity

T_d: $(n\times n)$ vector of unknown terms arising from unmodelled dynamics and external disturbances

q: $(n\times 1)$ vector of joint positions

$\dot{q}$: $(n\times 1)$ vector of joint velocities

$\ddot{q}$: $(n\times 1)$ vector of joint accelerations

In the computed-torque (also known as inverse model control) technique, a completely decoupled error dynamics equation can be obtained. The overall architecture of this model is shown in Figure 2.1. If the dynamic model is exact, the dynamic perturbations are exactly cancelled. The total torque driving the manipulator is given by:

$$T = M(q)u + \hat{V}(q,\dot{q}) + \hat{F}(\dot{q}) + \hat{G}(q) + T_f \qquad (2.2)$$

where u is defined as

$$u = \ddot{q}_d + K_v(\dot{q}_d - \dot{q}) + K_P(q_d - q) \qquad (2.3)$$

Substituting (2.2) into (2.1), and noting that $\hat{V} = V$, $\hat{F} = F$ and $\hat{G} = G$ because of the assumption of an exact dynamic model, we obtain:

$$M(q)[\ddot{q}_d + K_v(\dot{q}_d - \dot{q}) + K_P(q_d - q)] = 0 \qquad (2.4)$$

The following error vectors can be defined:

$$E = (q_d - q), \; \dot{E} = (\dot{q}_d - \dot{q}) \; \text{and} \; \ddot{E} = (\ddot{q}_d - \ddot{q}) \tag{2.5}$$

These are related by the following expression, since M(q) is positive definite:

$$\ddot{E} + K_v \dot{E} + K_p E = 0 \tag{2.6}$$

2.3 ADAPTIVE CONTROL OF ROBOT MANIPULATORS

As mentioned earlier, the computed-torque controller, powerful though it is, has performance problems because of its reliance on a fixed dynamic model. Robots have to face uncertainty in many dynamic parameters describing the dynamic properties of grasped loads. Sensitivity to such parameter uncertainty is especially severe in high-speed operations using direct-drive robots, for which no gear reduction is available to mask effective inertia variations. More generally, if advanced robots are designed to be capable of interacting with their environment in a precisely controlled way (e.g., providing accurate force or impedance control), they are likely to exhibit high sensitivity to external forces and load variations.

Two classes of approach have been actively studied to maintain performance in the presence of parameter uncertainties: robust control, and adaptive control. Because the adaptation mechanism keeps extracting parameter information in the course of operation, adaptive controllers can potentially provide consistent performance in the face of even very large load variations. They are therefore intuitively superior to robust controllers in this type of application.

The availability of effective adaptive controllers has been crucial in enabling the development of high-speed and high-precision robots. Two phases may be discerned in the history of adaptive robot control research, an approximation phase (1979-1985) and a linear-parameterization phase (1986-present). In the approximation phase, research relied on restrictive assumptions or approximations for adaptive-control design and analysis, for example linearization of robot dynamics, decoupling assumption for joint motions, or slow variation of the inertia matrix. Pioneering papers of this first phase include [1][2]. A reasonably complete review of these early methods is presented in [3].

The explicit introduction of the linear parameterization of robot dynamics in adaptive robotic control research represented a turning point inspired by earlier results developed in the context of parameter estimation [4][5]. Research in adaptive robot control could now fully account for the nonlinear, time-varying and coupled nature of robot dynamics, based on the possibility of selecting a proper set of equivalent parameters such that the manipulator dynamics depend linearly on these parameters. The resulting adaptive controllers can be classified into three categories: direct, indirect and composite. The direct adaptive controllers (e.g., [6]) use tracking errors of the joint motion to drive parameter adaptation. The indirect adaptive controllers [7][8][9], on the other hand, use prediction errors on the filtered joint torques to generate parameter estimates to be used in the control law. The composite adaptive controllers [10] use both tracking errors in the joint motion and prediction errors on the filtered torques to drive parameter estimation.

2.3.1 Direct adaptive controllers

Direct adaptive controllers use tracking errors of the joint motion to drive the parameter adaptation. In this class of adaptive controllers, the predominant concern of the adaptation law is to reduce the tracking errors.

An adaptive controller based on computed-torque control is proposed in [6][11], and global convergence is demonstrated. The authors' desire to maintain the structure of the computed-torque control scheme brings about the need to use acceleration measurements and to invert the estimated inertia matrix as part of the algorithm. In [12], the global tracking convergence of a new adaptive-plus-PD controller is established and its good performance is demonstrated experimentally. The algorithm only requires the system's state (joint positions and velocities) to be measured, and avoids inversion of the estimated inertia matrix by taking advantage of the inherent positive definiteness of the actual inertia matrix. Related work appears in [13][14][15].

2.3.2 Indirect adaptive controllers

Indirect adaptive-manipulator controllers, pioneered by Middleton and Goodwin [7], use prediction errors on the filtered joint torques to generate parameter estimates to be used in the control law. They show the global tracking convergence of their adaptive controller, which is composed of a modified computed-torque control and a modified least-squares estimator. The computation of the adaptive controller again requires inversion of the inertia matrix. The indirect adaptive controllers in [9] avoid this requirement by using a different modification of the computed-torque controller. Most of the indirect algorithms have to assume (or to develop procedures to guarantee) that the estimated inertia matrix remains positive definite in the course of adaptation. If only the load is to be estimated, a projection approach can be used to maintain this positive definiteness while preserving convergence properties, as the convexity result of [9] shows. Indirect controllers allow the vast parameter-estimation literature to be used to select time-variations of the adaptation gains.

2.3.3 Composite adaptive controllers

Composite adaptive controllers, studied in [10], use both tracking errors in the joint motions and prediction errors on the filtered torques to drive the parameter adaptation. They also allow standard parameter-estimation techniques to be exploited.

2.4 NEURAL-ADAPTIVE CONTROLLERS

Although conventional adaptive controllers are effective in compensating for the influence of structured uncertainty, it is not clear that adaptive means alone can overcome the effects of unstructured uncertainty. Where there is significant unstructured uncertainty in a robot system, some form of learning control is necessary if the controller is to be effective.

Both adaptive and learning control systems can be implemented using parameter adjustment algorithms, and both make use of performance feedback information gained through closed-loop interactions with the plant and its environment. Nevertheless, the goals of adaptation (i.e., updating behaviour through time) are distinct from those of learning (i.e., developing a static global model by associating behaviour with situation). A system that treats every distinct operating situation as novel is limited to adaptive operation, whereas a system that correlates past experiences with past situations, and that can recall and exploit those experiences, is capable of learning. Since a learning system must be capable of adjusting its memory to accommodate new experiences, a learning system must, in some sense, incorporate an adaptive capability. It should be noted, however, that the design and intended purpose of a learning system requires capabilities beyond that of adaptation.

A neural network with a learning algorithm called backpropagation was the first approach to adaptive control which incorporated a learning capability [16]. The neural network was used in a feedforward loop with a conventional feedback PD controller for manipulator control. As learning progressed, the feedforward/feedback compensation tended to move to the feedforward path with little feedback compensation. The inverse-dynamics model of the manipulator was considered to be obtained by the learned neural network. A later neural network controller described in [17][18] had many preprocessors computing various nonlinear transformations of input signals. The designer of this neural network controller was required to specify the structure of the controlled system in a fairly precise way and the effectiveness of the neural network controller in compensating for unstructured uncertainties was not clear. In [18], a three-layered neural network without the preprocessor was used, and the designer of the controller was not required to know the structure of the robotic manipulator. The actual trajectory of the manipulator followed the desired one well after the learning was finished, but when the desired trajectory was changed to one not used in the training of the neural network controller, the error between the desired trajectory and the actual one became large, and more learning was necessary. This means that the neural network had fitted a relationship between the input/output data but had not succeeded in learning the inverse-dynamics model: this has been a common observation with neural network controllers. Learning architectures for neural network controllers were proposed by Psaltis *et al.* [19]: these seemed to be more efficient in learning the nonlinear mechanical manipulator dynamics. However, for the general learning architecture, the data of the inputs and outputs of the plant have to be obtained by actually operating the plant in real time. This operation may take much time. Furthermore, although the neural network controllers were assumed to be effective in compensating for the unstructured uncertainties, no clear comparison of the tracking performances between the neural network controller and conventional adaptive schemes was made.

The incorporation of the idea of the computed-torque method within a nonlinear compensator using neural networks was proposed in [21]. Here, the neural networks were used to compensate for nonlinearities of the robotic manipulator rather than to learn the inverse dynamics. A comparison of the performance of the proposed neural network controller with that of the adaptive controller proposed by Craig *et al.* [11] was reported.

Another method for the direct control of robot manipulators using neural networks was proposed in [22]. In this, the control system consists of an inverse model of the robot dynamics which produces the forces/torques to be applied to the robot, given desired positions, velocities and accelerations, and a neural controller generating a correcting signal. This technique employs reinforcement learning for the updating of the network parameters.

Having reviewed these earlier attempts at neural-adaptive control, it is clear that improvements are desirable. A technique is required which provides both fast and robust on-line learning of the inverse dynamics of the robot. A new approach is to combine a neural network model-learning element with a servo PD feedback controller, as reported previously in [23]. The scheme resembles other feedforward control structures except that the manipulator's inverse-dynamics model is replaced by generic neural network models, one per joint. Each neural network model adaptively approximates the corresponding joint's inverse dynamics. A comparison of the tracking performances between the adaptive neural network controller and conventional adaptive schemes was reported in [23].

The first implementation of this new control structure used the backpropagation algorithm (BP) to update the neural network weights. When it first appeared, the BP algorithm represented a tremendous breakthrough in the control application of multilayer perceptrons, but its slow rate of convergence has since been recognised as one of the major drawbacks of the technique. Starting from a random initial state, the path to the global minimum is often strewn with local minima, causing oscillations around ravines in the weights space. Thus, although BP provides a robust technique for on-line learning of the inverse robot dynamics, the rate of convergence is too slow for this application because fast learning of the inverse dynamics is required. Hence, some means of improving the rate of convergence, and hence the speed of learning, is demanded

The common approach to speeding up the backpropagation algorithm is to copy various methods from the field of numerical analysis. For example, a modified version of the error backpropagation algorithm based on a conjugate gradient descent method can be used because this method has quadric convergence properties. However, a simpler approach is to avoid the BP algorithm altogether by considering the problem of adjusting the weights as a problem of estimating parameters. Then, well-known parametric system identification methods can be used, such as the recursive prediction error (RPE) method, the extended Kalman filter (EKF) approach, and a decomposition method using the QR-WRLS algorithm. These are discussed further in Section 2.6.

2.5 PARAMETRIC SYSTEM IDENTIFICATION METHODS

It is pertinent, therefore, to review some of the fundamental properties of parametric system identification schemes. A convenient starting point is to review those schemes which are based on linear system models [24][25] and then to consider which of these can be adopted in nonlinear system identification schemes.

Consider a discrete-time, dynamic system which can be represented by a normalized, deterministic, auto-regressive, moving-average (DARMA) model:

$$y(k+1) + a_0 y(k) + \ldots + a_{n-1} y(k-n+1) = b_0 u(k) + b_1 u(k-1) + \ldots + b_l u(k-l) \quad (2.7)$$

This equation can be expressed in a parametric form known as the *regressor model*:

$$y(k+1) = \Phi^T(k)\Theta_0 \quad (2.8)$$

where

$$\Phi^T(k) = [y(k),\ y(k-1),\ \ldots,\ y(k-n+1),\ u(k),\ \ldots,\ u(k-l)]^T \quad (2.9)$$

is the regressor vector and

$$\Theta_0 = \left[-a_0,\ -a_1,\ \ldots,\ -a_{n-1},\ b_0,\ \ldots,\ b_l\right]^T \quad (2.10)$$

is the parameter vector. The system dynamics in this form can be referred to as *the predictor form*.

The next step is to estimate the parameter vector Θ_0 using the real time operational input/output data of the system. A generic system identification scheme is shown in Figure 2.2.

The output of the adaptive estimation model can be written as

$$\hat{y}(k+1) = \Phi^T \hat{\Theta}(k) \quad (2.11)$$

where $\hat{y}(k+1)$ represents the estimated system output, and $\hat{\Theta}(k)$ stands for the estimated parametric vector.

There are two main methods used for parametric estimation [24]:

(a)   equation error method,
(b)   output error method.

In the equation error method, error $e(k+1)$ is defined to be the error in the estimated model equation, that is,

$$e(k+1) = y(k+1) - \Phi^T(k)\hat{\Theta}(k) \quad (2.12)$$

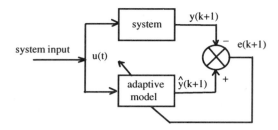

Figure 2.2 A generic system identification scheme

where the regressor vector $\Phi^T(k)$ is defined in terms of the states of the actual system.

In the output error method the regressor vector $\Phi^T(k)$ is defined in terms of the estimated model, that is,

$$\Phi^T(k) = \left[\hat{y}(k), \, ..., \hat{y}(k-n+1), \, u(k), \, ..., \, u(k-l) \,\right] \tag{2.13}$$

The main difference between these two schemes is that the estimated system output $\hat{y}(k+1)$ used in the equation error method is based on the regression vector of the original system and not on the estimated model. Parameter convergence properties of the equation error method are far better than those of the output error method [24]. The equation error method will therefore be used in the proposed system identification scheme, assuming that the previous values of the system output are available.

Based on equations (2.11) and (2.12), we denote the parameter error vector as

$$\tilde{\Theta}(k) = \Theta_0 \, - \, \hat{\Theta}(k) \tag{2.14}$$

The output error vector can be defined as:

$$e(k+1) = y(k+1) - \hat{y}(k+1) \tag{2.15}$$
$$= y(k+1) - \Phi^T(k)\hat{\Theta}(k)$$
$$= \Phi^T(k)\Theta_0 - \Phi^T(k)\hat{\Theta}(k)$$
$$= \Phi^T(k)\tilde{\Theta}(k)$$

The next step is to construct an algorithm which makes use of the error term $e(k+1)$ to estimate the true values of the parameters. There are various globally convergent parameter update algorithms which have different merits. The best known are:

(a)   backpropagation algorithm (BP),
(b)   recursive least square (RLS),
(c)   recursive prediction error (RPE).

Two important properties frequently used in adaptive system theory are:

(i)   **Persistent excitation conditions.** Conditions under which good parameter estimates can be obtained are usually defined as the persistency of excitation conditions. This means that the input sequence driving the system should be sufficiently rich in its frequency content that various modes of the system can be excited.

(ii)   **Certainty equivalence principle.** It is basically an *a priori* assumption that the estimated parameters are close to their true values, so that the estimated model can be used in a control scheme as the actual dynamic model of the system under study.

Our main interest in this work is a nonlinear system for which parametric models are not available. However, we assume some *a priori* knowledge about the structural properties of the system, like the system order, invertibility and observability.

2.5.1 Nonlinear system identification based on ANN models

For the identification of nonlinear systems, an artificial neural network approach is attractive. This section discusses such a novel system identification scheme based on an ANN model of the system under study. The *a priori* assumption here is that a parametric model of the system dynamics is not available. However, it is assumed that certain structural properties are given.

Consider the state space form of the system defined by

$$x(k+1) = \Phi(x(k),\ u(k))\ ;\quad y(k) = Y(x(k)) \tag{2.16}$$

The following assumptions about the system are stated before devising the identification scheme.

(a)   **Assumption 1.** The order of the system (the dimension of the state vector) is known a priori.

(b)   **Assumption 2.** System states are measurable.

Given the state space representation above of the system, the output at the $(k+1)$th instant can be written as

$$\begin{aligned} y(k+1) &= Y(x(k+1)) \\ &= Y(\Phi(x(k),\ u(k))) \end{aligned} \tag{2.17}$$

Following the same procedure:

$$y(k+2) = Y(\Phi\ (\Phi(x(k),\ u(k)),\ u(k+1))) = Y_2(.)$$
$$..\quad =\quad ...$$
$$..\quad =\quad ... \tag{2.18}$$
$$..\quad =\quad ...$$
$$y(k+n) = Y\ (\Phi...\ (\Phi(x(k),\ u(k)),\ u(k+1)),\ u(k+n-1))$$

Thus

$$y(k+n) = Y_n(x(k),\ u(k),\ ...,\ u(k+n-1)) \tag{2.19}$$

The above equations provide us with n nonlinear equations obtained by observing the sequences $[y(k+1),\ ...,y(k+n)]^T$ and $[u(k+1),...,u(k+n-1)]^T$. The unknown state vector $x(k)$ is common to all n equations and is in fact the solution of this system of equations. Therefore the condition that we assume holds is that, given the above set of nonlinear equations, there exists a unique solution $x(k)$ of the system. The state observability assumption is also related to the invertibility condition of the system under study. Referring to equation (2.19) let

$$u(k) = [u(k),\ ...,\ u(k+n-1)] \tag{2.20}$$

denote the control vector. Given the output $y(k+n)$, if the system is invertible, the control vector $U(k)$ can be solved as a function of $y(k+n)$ and $x(k)$ [25], that is,

$$U(k) = Y_n^{-1}\ (y(k+n),\ ...,\ x(k)) \tag{2.21}$$

Clearly, the inversion requires information on the state vector $x(k)$. Equation (2.21) is the fundamental relation representing the inverse dynamics of the nonlinear system.

Based on the above assumptions, a nonlinear system identification scheme utilizing the computational properties of ANN can now be presented. This identification scheme makes use of a multilayer ANN as the adaptive model of the nonlinear system under study.

2.5.2 Mathematical analysis

A number of research papers which have appeared in different journals give mathematical proofs showing that multilayer ANNs can be used as functional approximators [26][27][28]. In the following, we quote an existence theorem on the functional approximation property of ANNs. The theorem has been studied in detail by Funahashi [26]. This theorem shows that any arbitrary mapping from Γ^n to Γ^r can be approximated by a network with at least three layers. The above results are quite powerful in showing the existence of nonlinear mapping capabilities in neural networks. Simple, nonlinear, monotonic increasing functions are assumed as activations of the nodes; hence this justifies the use of sigmoidal nonlinearities.

The main point so far is that the above results supply us with strong mathematical justification for utilizing neural networks in nonlinear adaptive signal processing and control applications. In the rest of this work, we base our analysis of the proposed control architectures on the above theorem. Computation of the desired weights for achieving a certain mapping is achieved by iteratively updating the weights based on the minimization of the error criterion.

2.6 ESTIMATION OF WEIGHT VECTOR

Given some knowledge about a particular problem in the form of input variables and outputs, it is desirable to estimate the weights vector in such a way that, when the system is presented with a new set of features, it can predict the correct outcome. In other words we wish to represent the knowledge relating the features to the outcomes by the weight vector.

Let $\Theta = [\theta_1,...,\theta_n]^T$ represent all the unknown weights and thresholds of the network. A network with a single layer of hidden units can then be represented by the model:

$$\hat{y}_i(k,\theta) = \sum_{j=1}^{j=n_1} \omega_{ij}^2 x_j^1(k) = \sum_{j=1}^{j=n_1} \omega_{ij}^2 g\left[\sum_{m=1}^{m=n_o} \omega_{jm}^1 x_m(k) + b_j^1\right] \quad \text{with } 1 \le i \le n_1$$

(2.22)

where $x(k) = [x_1(k), ..., x_{n_0}(k)]^T$ is the input vector to the network, ω_{jm} are the weights, b_j are the thresholds and g[.] is the node activation function.

Many algorithms for estimating the weight vector are now available. These include recursive prediction error (RPE), recursive least squares (RLS) and extended Kalman filter (EKF). All of these existing algorithms have deficiencies in various respects for the application in question, and hence a new and better algorithm is also introduced in the discussion below, known as parallel weighted recursive least squares (PWRLS). This can be implemented in various forms, which are compared in terms of their performance.

This latter algorithm may be viewed as the deterministic counterpart of the Kalman filter theory [29]. The structure and performance of these various weight estimation algorithms are compared below. However, the EKF algorithm has been omitted from this discussion because it is very similar to the RLS algorithm, having almost the same basic mathematical structure.

2.6.1 The recursive prediction error (RPE) algorithm
The equations for the RPE algorithm are (Ljung and Soderstrom 1983)[30]

$$e(k) = y(k) - \hat{y}(k) \tag{2.23}$$

$$P(k) = \frac{1}{\lambda(k)}\left[P(k-1) - P(k-1)\Psi(k)\lambda(k)I + \Psi^T(k)P(k-1)\Psi(k)\right]^{-1}\Psi^T(k)P(k+1) \tag{2.24}$$

$$\hat{\Theta}(k) = \hat{\Theta}(k-1) + P(k)\Psi(k)e(k) \tag{2.25}$$

where

$\Psi(k)$ is a gradient vector.

$P(k)$ is a covariance matrix.

$\hat{\Theta}(k)$ are parameters to be estimated

$e(k)$ is the prediction error

$\lambda(k)$ is the forgetting factor

$$\begin{cases} \lambda(k) \prec 1 & \text{at the initial stage} \\ \lambda(k) \to 1 & \text{as } t \to \infty \end{cases}$$

and then $\lambda(k)$ will be in the form

$$\lambda(k) = \lambda_0 \lambda(k-1) + (1-\lambda_0) \tag{2.26}$$

The number of parameters to be estimated with a single hidden layer is given by

$$n_\theta = (n_0 + 1)n_1 + n_1 n_2 \tag{2.27}$$

The elements of the gradient $\Psi(k,\Theta)$ can be obtained by differentiating (2.22) with respect to Θ_i, Billings *et al.* [31].

$$\Psi_{ij} = \frac{d\,\hat{y}_j}{d\,\Theta_i} = \begin{cases} x_m^1 & \text{if } \Theta_i = \omega_{jm}^2 & 1 \le m \le n_1 \\ x_m^1(1-x_m^1)\omega_{jm}^2 & \text{if } \Theta_i = b_m^1 & 1 \le m \le n_1 \\ x_m^1(1-x_m^1)\omega_{jm}^2 x_m & \text{if } \Theta_i = \omega_{ml}^1 & 1 \le m \le n_1 \\ 0 & \text{otherwise} & 1 \le 1 \le n_0 \end{cases} \tag{2.28}$$

Computational efficiency can be improved by converting the algorithm into the parallel form given below, known as the parallel recursive prediction error algorithm.

2.6.1.1 Parallel recursive prediction error algorithm (PRPE)

The PRPE algorithm is derived from the conventional RPE algorithm by choosing the Hessian $\overline{H}(\Theta)$ to be a near-diagonal matrix, Chen *et al.* [32]. The equations for the PRPE will then be as follows:

$$e(k) = y(k) - \hat{y}(k) \tag{2.29}$$

$$P_i(k) = \frac{1}{\lambda(k)} \left[P_i(k-1) - P_i(k-1)\Psi_i(k)\lambda(k)I + \Psi_i^T(k)P_i(k-1)\Psi_i(k) \right]^{-1} \Psi_i^T(k)P_i(k-1)$$

$$\tag{2.30}$$

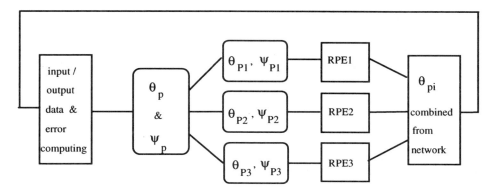

Figure 2.3 Implementation of PRPE algorithm

$$\hat{\Theta}_i(k) = \hat{\Theta}_i(k-1) + P_i(k)\Psi_i(k)e(k) \qquad (2.31)$$

This algorithm can be viewed as applying the conventional RPE algorithm to each neuron in the network. The implementation of the PRPE algorithm is summarized in Figure 2.3.

2.6.2 Recursive least squares algorithm (RLS)

Using the RLS, the learning algorithm is purely linear in the sense that the matrix vector equation to be solved has been transformed into the equation of a linear system. Accordingly, this linear algorithm and its potential variants will benefit from the well-understood theoretical properties of RLS and VLSI architecture.

In principle, once the linearization is achieved at iteration (i) and pattern (n), any LMS-type algorithm can be employed to update the weight estimates. The conventional WRLS algorithm takes the form of two recursions in the case of node (k) in the output layer:

$$P(i,n) = I - \frac{P(i,n-1)\,\bar{y}'_k(n)\left[\bar{y}'_k(n)\right]^T P(i,n-1)}{1+\left[\bar{y}'_k(n)\right]^T P(i,n-1)\bar{y}'_k(n)} \qquad (2.32)$$

and

$$\omega_k(i,n) = \omega_k(i,n-1) + P(i,n)\,\bar{y}'(n)\left[\bar{t}_k(n) - y_k(n)\right] \qquad (2.33)$$

$\omega_k(i,n)$ is the estimate of the weights ω_k following pattern (n) in the ith iteration through the training data, and $P^{-1}(i,n)$ is the covariance matrix at the same time, given by

$$P^{-1}(i,n) = \sum_{j=1}^{n} \bar{y}'_{k(n)}\left[\bar{y}'_k(n)\right]^T \qquad (2.34)$$

2.6.3 Parallel weighted recursive least squares algorithm (PWRLS)

The rate of weight convergence in estimating the weight vector is influenced greatly by the number of weights which are adjusted at each iteration of the algorithm. For this reason, we have developed a new algorithm called parallel weighted recursive least squares (PWRLS), which updates a large number of weights at once.

All training algorithms calculate the weights using several iterations through the training data. Gradient descent algorithms form a linear approximation to the error surface E at the present weights, and change the weights in the direction which reduces the total error. The error is a function of all the weights. The BP algorithm uses only one weight at a time for this approximation and is consequently slow to converge. To improve the rate of convergence, newer techniques (RPE, PRPE) use more weights simultaneously, typically all the weights connected to one node, to form an approximation to E at each iteration. Our improved algorithm, PWRLS, forms an approximation to E by updating an even greater number of weights simultaneously. Various possible implementations of this concept are considered below, and a comparison of their relative performance follows.

2.6.3.1 Parallel-node WRLS (PN-WRLS)

If only the weights connected to output node (k) in the output layer are allowed to change (Figure 2.4) with the others held constant, minimizing E is then minimizing E_k, where:

$$E_k = \sum_{n=1}^{N} \left[t_k(n) - f(\omega_k^T y'(n)) \right]^2 \tag{2.35}$$

The optimal solution, say ω_k^*, to this problem is simpler if the modelled output depends only upon a linear combination of the inputs using pattern invariant (constant) weights. In the linear cases:

$$y_k(n) = f(\omega_k \, y'(n)) = \beta \omega_k \, y'(n) \tag{2.36}$$

for some β (where β is constant, and can be taken to be unity if required), and the error expression takes the form:

$$E_k = \sum_{n=1}^{N} \left[t_k(n) - (\omega_k^T y'(n)) \right]^2 \tag{2.37}$$

The solution is a classical LS normal equation [33].

The linearization techniques adapted in this work can be explained in terms of error surface analysis. Suppose we wish to construct a linearized error surface $\overline{E}_k$ (similar

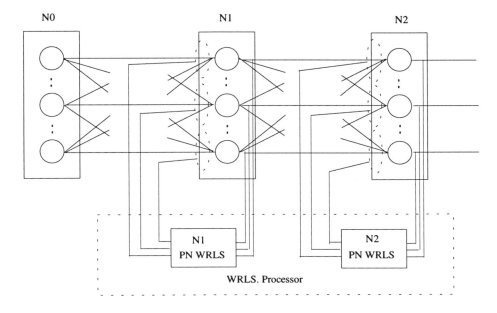

Figure 2.4 Implementation of the PN-WRLS algorithm

in some sense to E_k in a neighbourhood of the present weights). The objective is to find, by means of a linear algorithm, a close approximation of ω_k to the local (ideally it will be the global) minimum of E_k, say ω^*.

The algorithm to be described proceeds in iterations, indexed by (i=1,2,...). Each iteration represents one complete training cycle through the N patterns. By modifying the data pairs, we create a linear error surface which is similar to the nonlinear one in the neighbourhood of $\omega_k(i-1)$. We can then find the ω_k which minimize $\overline{E}_k$ by simple linear LS. The similarity follows from two criteria:

$$(1)\quad \overline{E}_k\, \omega_k(i-1)\left\{\overline{y}\,'_k(n),\ \overline{t}_k(n),\ n\in[1,N]\right\} \tag{2.38}$$

$$= \overline{E}_k\, \omega_k(i-1)\left\{y'_k(n),\ t_k(n),\ n\in[1,N]\right\}$$

$$(2)\quad \left.\frac{\partial \overline{E}_k}{\partial \omega_k}\right|_{\omega_k\,=\,\omega_k(i-1)} = \left.\frac{\partial E_k}{\partial \omega_k}\right|_{\omega_k\,=\,\omega_k(i-1)} \tag{2.39}$$

The first task is to manipulate the pairs $\{(\bar{y}'_k(n), \bar{\imath}_k(n)), n \in [1, N]\}$ so that these criteria hold. When criterion (1) is met for each (n), we take the partial derivatives required in criterion (2). By modifying the data pairs, we have created a linear error surface which is similar to the nonlinear one in the neighbourhood of $\omega_k(i-1)$. In turn, this implies convergence to the nonlinear solution for the weights, using purely linear techniques.

2.6.3.2 Parallel layer weighted recursive least squares (PL-WRLS)

The above linearization algorithm to update the network weights can be extended to update all the weights of one layer simultaneously, using a structure as indicated in Figure 2.5. The linearized error surface $\bar{E}$ can be constructed by replacing the nonlinearity $f(.)$ by a linear approximation.

Suppose we wish to update weights in the output layer simultaneously. We must linearize all output nodes. For node k in the output layer, the output in response to input n is computed as

$$y_k(n) = f(u) = f\left(\sum_{j=1}^{N_1} \omega_{kj} y'\right) \tag{2.40}$$

$(j \in [1, N_1]$: nodes in hidden layer)

The output of node k after $f(u_k(n))$ has been replaced by the derivative $\dot{f}(u_k(n)) = K_k u_k(n) + b_k$ and $\bar{y}_k(n)$ output of node k is given by

$$\bar{y}_k(n) = K_k(n)\left[\sum_{j=1}^{N_1} \omega_{kj} y'_j(n)\right] + b_k(n) \tag{2.41}$$

or:

$$\bar{z}_k(n) = K_k(n)\left[\sum_{j=1}^{N_1} \omega_{kj} y'_j(n)\right] \tag{2.42}$$

with:

$$\bar{z}_k(n) = \bar{y}_k(n) + b_k(n) \tag{2.43}$$

Since $\bar{y}(n) = Y_k(n)$ at the present weights, the error at the kth node will be the same for the linearized and original network if the target value for $\bar{z}_k(n)$, say $\bar{\imath}_k(n)$, is taken to be

$$\bar{\imath}_k(n) = t_k(n) - b_k(n) \tag{2.44}$$

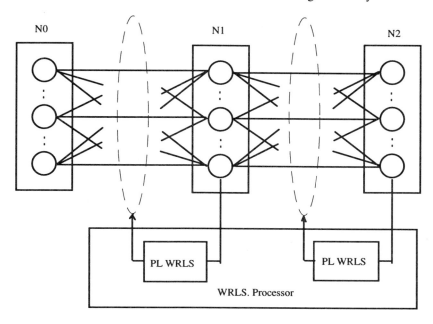

Figure 2.5 Implementation of PL-WRLS algorithm

and the linearized inputs to node k at pattern n are

$$\bar{x}_{kj}(n) = K_k(n)y'_j(n) \qquad \left(j \in [1, N_1] \right) \tag{2.45}$$

The problem has effectively been reduced to one of estimating weights for a single layer linear network.

In order to update all the weights in the output layer simultaneously, the system of $(N \times N_1)$ equations:

$$\bar{t}_k(n) = \sum_{j=1}^{N_1} \bar{x}_{kj}(n) \overline{\omega}_{kj} \tag{2.46}$$

where

$j \in [1, N_1]$, hidden layer nodes

$k \in [1, N_2]$, output layer nodes

$n \in [1, N]$, number of patterns

must be solved for the LS estimate of the $N_1 \times N_2$ weights ω_{kj}. However, since all weights in the hidden layer are fixed, the outputs y_j' are independent of k. This means that equations indexed by different values of k are independent of one another and the sets of weights connected to different outputs may be updated independently. In the output layer, therefore, there is no theoretical difference between PL-WRLS and PN-WRLS.

In the hidden layer, the equation:

$$\bar{t}_k(n) = \sum_{j=1}^{N_1} \sum_{p=1}^{N_0} \bar{x}'_{kjp}(n) \bar{\omega}_{jp}, \quad (p \in [1, N_0], \text{ number of inputs}) \tag{2.47}$$

Unlike for the output layer, the problem cannot be decomposed into separate solutions for sets of weights connected to individual nodes in the hidden layer. This is a reflection of the fact that all weights in the hidden layer are coupled through their mixing in the output layer. This means that the simultaneous solution for all weights in the hidden layer should be beneficial with respect to a PN-WRLS. The updating procedure is repeated until convergence.

The algorithm proceeds as follows:

(i)   Linearize the system around the present weights.
(ii) Change the weights by a small amount to decrease error.
(iii) Repeat until convergence.

2.6.3.3 Parallel network WRLS (PNet-WRLS)
For the same reason that PL-WRLS estimation of weights is beneficial, we should expect even more benefit from complete network updating if such were possible, as indicated in Figure 2.6. In the case of a single node in the output layer of the network ($k=1$):

$$\omega_{lj}^+ = \omega_{kj} \omega'_{jl} = \omega_{1j} \omega'_{jl} \tag{2.48}$$

$$\bar{y}_l(n) - b_l(n) = \sum_{j=1}^{N_1} \sum_{l=1}^{N_0} \left[K_l(n) K_l'(n) x_l(n) \right] \omega_{jl}^+ + \sum_{j=1}^{N_1} \left[K_l(n) b'_j \, \omega_{lj} \right] \tag{2.49}$$

This can be interpreted as an attempt to train one single linear layer with one output and $(N_0 \times N_1) + N_1$ inputs. In this case there will only be N linearized training patterns.

Note that $\omega_k(i,0) = \omega_k(i-1,N)$ and similarly for the covariance matrix. This leads on to the problem of initialization. Clearly, some means of initializing the weights $\omega(i,0)$ and covariance $P(0,0)$ is required. Unfortunately, the WRLS algorithm, in common with the RPE and EKF algorithms, fails to answer this question. The inverse covariance matrix   contains   theoretically   infinite   values   at   the   outset   and   practical

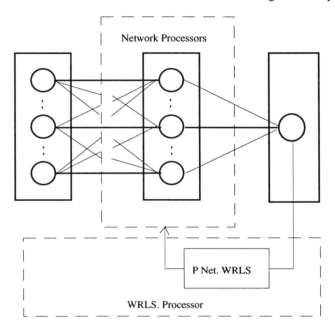

Figure 2.6 Implementation of PNet-WRLS algorithm

guidance about proper initialization of the weights is not available. This can lead to convergence difficulties, even though the convergence ability is generally better than for the alternative algorithms already discussed.

A further problem can arise in respect of the forgetting factor used in the WRLS algorithms, which is equivalent to using a weighted error criterion with time-varying weights. Performance is critically dependent on correct choice of this forgetting factor. If the forgetting factor is large, convergence is slow. If it is small, then past values are forgotten more quickly, but the algorithm may have convergence problems. To avoid such performance problems in these circumstances, a further development of the WRLS algorithm has been made in which QR decomposition is applied in a method labelled QR-WRLS .

2.6.4 The QR-WRLS algorithm
The QR-WRLS algorithm has distinct advantages over conventional RPE and RLS (EKF) algorithms in the following respects:

(a)   It does not suffer from the initialization problems noted above.
(b)   It is robust numerically, as no matrix inversions are done.
(c)   It permits the inclusion of several very flexible forgetting strategies.

The derivation of the QR-WRLS algorithm is given below.

2.6.4.1   *Weight estimation using QR decomposition*

In this approach, the decomposition problem is reformulated as follows. Given a matrix A, find an orthogonal matrix Q such that

$$Q A = R \qquad\qquad (2.50)$$

is an upper triangular matrix. The matrix Q is expressed as a product of simple orthogonal matrices called Givens rotations.

The weights ω_k in the output layer of three-layer neural networks can be estimated using a weighted LS algorithm (WLS) as follows.

Define a vector $z(t)$ at time t as

$$z(t) = \left[z_1(t), \ ...., \ z_{N_1} \right] \qquad\qquad (2.51)$$

where $z_i(t)$ are the outputs from the hidden nodes and N_1 are the number of hidden nodes.

Define the matrix $X(t)$ and an output vector, $Y(t)$ at time t as

$$X(t) = \begin{bmatrix} z(1) \\ z(2) \\ .. \\ .. \\ z(t) \end{bmatrix} \qquad \text{and} \qquad Y(t) = \left[y(1), \ y(2), \ ...., \ y(t) \right] \qquad\qquad (2.52)$$

The normal equation can be written as

$$Y(t) = \Theta(t) X(t) \qquad \text{and} \qquad \Theta(t) = \left[\omega_1(t), \ ......, \ \omega_{N_1}(t) \right]^T \qquad\qquad (2.53)$$

The QR-WRLS algorithm method is based upon transforming this system into an upper triangular system by applying a series of orthogonal operators (Givens rotations). The resulting system is as follows:

$$\begin{bmatrix} T(i, N) \\ ----- \\ O_{(N-N_1) \times N_1} \end{bmatrix} \omega_k(i) = \begin{bmatrix} d_1(i, N) \\ ---- \\ d_2(i, N) \end{bmatrix} \qquad\qquad (2.54)$$

where the matrix $T(i,n)$ is $(N_1 \times N_1)$ upper triangular, and $O_{(N-N_1) \times N_1}$ denotes the $(N-N_1) \times N_1$ zero matrix. The solution for $\omega_k(i)$ is obtained by backsubstitution (Figure 2.7).

Weight estimation using the recursive QR-WRLS algorithm proceeds as follows.

(a)   *Initialization.* Initialize an $(N_1 \times 1) \times (N_1 \times 1)$ working matrix A to a null matrix.

(b)   *Recursion.* For $i = 1,2,....$(iteration); and for $n = 1,2,...,N$ (pattern),

    (i)   Enter the next equation into the bottom row of A :

$$\left[\left[\bar{y}_k (n) \right]^T \;\middle|\; \bar{t}_k(n) \right]$$

    (ii)   Rotate the new equation into the system using

$$A'_{mk} = A_{mk}\sigma + (A_{N_1+1})\tau s$$
$$A'_{N_1+1,k} = -A_{mk}\tau s + (A_{N_1+1,k})\sigma s$$
$$\text{for } k \in \left[m, N_1 + 1 \right] \text{ and } m \in \left[1, N_1 \right]$$

where

$$s = \text{unity}$$
$$\sigma = A_{mm}/\rho$$
$$\tau = A_{N_1+1,m}/\rho$$
$$\rho = (A_{mm}^2 + A_{N_1+1,m}^2)^{1/2}$$
$$A_{mk} \quad \text{is the } (m,k) \text{ element of } A \text{ pre-rotation}$$
$$A'_{mk} \quad \text{is the } (m,k) \text{ element of } A \text{ post-rotation}$$

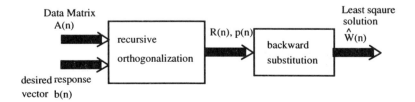

Figure 2.7 Block diagramatic representation of the QR-WRLS

(iii)   Solve for the least squares estimate of the weights ω_k if desired [solution after the nth pattern will produce what has been called $\omega_k(i, N)$ in the text].

(iv)   If $n \prec N$, increment n. Otherwise check convergence criterion and increment i and reset n if not met.

Termination. Stop when some convergence criterion is met.

The A matrix is defined as four partitions. Following the rotation of the nth equation, in step 2:

$$
A = \begin{bmatrix} T(i,n) & | & d_1(i,n) \\ -------- \\ O_{1 \times N_1} & | & d_2(i,n) \end{bmatrix}
\tag{2.55}
$$

2.6.4.2 On-line learning using QR-WRLS algorithm

The QR-WRLS learning algorithm is as follows.
1.   Initialize the weights of the network.
2.   Weight updating
   (i) present input pattern to the system,
   (ii) linearized training patterns are calculated for the last layer nodes,
   (iii) rotate these patterns into the corresponding A matrix,
   (iv) the target outputs are calculated,
   (v) update the A matrices.
3.   A new training pattern is presented to calculate a new set of linearized inputs/outputs.
4.   Repeat steps 2-3 until all the training patterns are used.
5.   Using the A matrices, update the weights.
6.   Repeat steps 2-5 until convergence.

2.6.4.3 Data reduction algorithm

At each iteration, the number of training patterns used can be reduced. As the training proceeds, a few of the linearized training patterns become dominant, and the weight change is largely dependent on these training patterns. Although the original training patterns do not change, the linearized training patterns are dependent on the present weights and so the linearized training patterns generally change at each iteration. For this reason, at one iteration a linearized training pattern may have a large effect on the weight change, while at another iteration, after the weights have changed, the linearized training pattern may have a negligible effect. The goal is to be able to determine which training patterns will have a very small effect on the resulting weights and avoid the rotation of these into the system of equations, thereby reducing the number of computations at this iteration and making training more efficient.

2.7 APPLYING ANN-BASED CONTROL TO ROBOT MANIPULATORS

The design of current industrial robots is based on simple joint servo-mechanisms assigned to the different joints of the arm. This results in reduced servo-response speed, thus limiting the precision and accuracy of the end effector and producing sub-optimal performance. The increasingly sophisticated tasks required of robot manipulators have called for better control techniques to enhance high-speed tracking accuracy whilst operating in uncertain environments. Many control schemes have been developed to overcome this problem. However, the performance of all these methods is highly dependent on the accurate modelling of robot dynamics. The advantage of a neural-adaptive control technique for robot manipulators is that it avoids this *a priori* modelling problem. The control technique is generic in the sense that the controller parameters are not dependent on any parameter estimation, as opposed to many conventional adaptive control methods.

Various mechanisms can be considered for incorporating neural networks into a standard computed-torque controller. An effective method using two neural networks was claimed to be effective in [34]. These networks are installed in place of the estimated models $\hat{M}$, $\hat{V}$. One of the networks identifies the inertia matrix M and the other one identifies the centrifugal/Coriolis V and such unstructured uncertainties as friction forces F. But the neural networks in this method have to learn not only the uncertainties but also the structure of the robotic manipulator. Thus, the adaptive control architecture that will be presented here maintains the same structure as the computed-torque servo, but in addition has an adaptive element that estimates the true values of the unknown parameters [35].

For an n-link rigid manipulator the vector dynamic equation is given by (from equations (2.2) and (2.3)):

$$T = M(q)\ddot{q} + V(q,\dot{q}) + F(\dot{q}) + G(q) \tag{2.56}$$

Although the equations of motion (2.56) are complex and nonlinear for all but simple robots, they have several fundamental properties which can be exploited to facilitate control system design.

(a) **Property 1**. The inertia matrix $M(q)$ is symmetric, positive definite and both $M(q)$ and $M^{-1}(q)$ are uniformly bounded as a function (q) of R^n.

(b) **Property 2**. There is an independent control input for each degree of freedom.

(c) **Property 3**. The Euler-Lagrange equation for the robot is linear in the unknown parameters.

All the unknown parameters are constant (e.g. link masses, link lengths, moments of inertia, etc.) and appear as coefficients of known functions of the generalized coordinates. By defining each coefficient or a linear combination of them as a separate parameter, a linear relationship results and equation (2.56) can be rewritten as: [35][4][5].

$$T = Y(q,\dot{q},\ddot{q})\ P \tag{2.57}$$

where $Y(q,\dot{q},\ddot{q})$ is an $(n \times r)^T$ matrix of known functions which is referred to as the regressor matrix and $P = [p_1, p_2, ..., p_r]^T$ is an r-dimensional vector of parameters.

Consider a two-link manipulator whose dynamic equations are given by equation (2.56) and:

$$M(q) = \begin{bmatrix} l_1^2(m_1 + m_2) + m_2 l_2^2 + 2m_2 l_1 l_2 c_2 & m_2 l_2^2 + m_2 l_1 l_2 c_2 \\ m_2 l_2^2 + m_2 l_1 l_2 c_2 & m_2 l_2^2 \end{bmatrix} = \begin{bmatrix} M_{11} & M_{12} \\ M_{21} & M_{22} \end{bmatrix} \tag{2.58}$$

$$V(q,\dot{q}) = \begin{bmatrix} -2m_2 l_1 l_2 s_2 \dot{q}_1 \dot{q}_2 - m_2 l_1 l_2 s_2 \dot{q}_2^2 \\ m_2 l_2 s_2 \dot{q}_1^2 \end{bmatrix} \tag{2.59}$$

$$G(q) = \begin{bmatrix} (m_1 + m_2)gl_1 c_1 + m_2 gl_2 c_{12} \\ m_2 gl_2 c_{12} \end{bmatrix} \tag{2.60}$$

$$F(\dot{q}) = \begin{bmatrix} v_1 \dot{q}_1 \\ v_2 \dot{q}_2 \end{bmatrix} \tag{2.61}$$

with:

$$c_i = \cos(q_i)$$
$$s_i = \sin(q_i)$$
$$c_{12} = \cos(q_1 + q_2)$$
$$v_i = \text{viscous friction for joint } (i).$$

Defining the parameter vector P in equation (2.57) as,

$$\begin{bmatrix} p_1 & p_2 & p_3 & p_4 \end{bmatrix}^T = \begin{bmatrix} m_1 & m_2 & m_3 & m_4 \end{bmatrix}^T \tag{2.62}$$

the torque in (2.57) can then be written as

$$T = Y(q,\dot{q},\ddot{q})P = \begin{bmatrix} y_{11} & y_{12} & y_{13} & 0 \\ 0 & y_{22} & 0 & y_{24} \end{bmatrix} \begin{bmatrix} p_1 \\ p_2 \\ p_3 \\ p_4 \end{bmatrix} \tag{2.63}$$

The control law of the computed-torque method is expressed as follows:

$$T = \hat{M}(q)\ddot{q} + \hat{V}(q,\dot{q}) + \hat{F}(\dot{q}) + G(q) \tag{2.64}$$

where $\hat{M}(q)$, $\hat{V}(q,\dot{q})$, $\hat{F}(\dot{q})$, $\hat{G}(q)$ denote the estimates of $M(q)$, $V(q,\dot{q})$, $F(\dot{q})$ $G(q)$ respectively and:

$$u = \ddot{q}_d + K_v(\dot{q}_d - \dot{q}) + K_p(q_d - q) = \ddot{q}_d + K_v\dot{E} + K_p E \tag{2.65}$$

Thus, estimated robot dynamics can be written as

$$\hat{M}(q)\ddot{q} + \hat{V}(q,\dot{q}) + \hat{F}(\dot{q}) + G(q) = Y(q,\dot{q},\ddot{q})\hat{P}, \text{ with } \hat{P} = [\hat{p}_1,...,\hat{p}_r]^T \tag{2.66}$$

Substituting the control vector (2.66) into robot dynamics equations (2.56) we get

$$\hat{M}(q)u + \hat{V}(q,\dot{q}) + \hat{F}(\dot{q}) + \hat{G}(q) = M(q)\ddot{q} + V(q,\dot{q}) + F(\dot{q}) + G(q) \tag{2.67}$$

From (2.67),

$$\tilde{M}(q)\left[\ddot{E} + K_v\dot{E} + K_p E\right] = \tilde{M}(q)a + \tilde{V}(q,\dot{q}) + \tilde{F}(\dot{q}) + \tilde{G}(q) \tag{2.68}$$

$$\tilde{M}(q)\left[\ddot{E} + K_v\dot{E} + K_p E\right] = Y(q,\dot{q},\ddot{q})\tilde{P}$$

where $\tilde{M} = M(q) - \hat{M}(q)$ and so on $(\tilde{\cdot}) = (.) - (\hat{\cdot})$; $(\tilde{P} = P - \hat{P})$ is the parameter error vector with the assumption $\tilde{M}(q)$ is nonsingular.

$$\left[\ddot{E} + K_v\dot{E} + K_p E\right] = \tilde{M}^{-1}(q) \, Y(q,\dot{q},\ddot{q})\tilde{P} = \Phi(q,\dot{q},\ddot{q})\tilde{P} \tag{2.69}$$

Equation (2.69) is a decoupled linear system driven by a nonlinear function.

The proposed neural network controller is shown in Figure 2.8. The ANN model is used to model the inverse-dynamics structure of each joint for nonlinear compensation of the manipulator. The structure of the controller is simple and it is easy to compare the performance with conventional adaptive controllers.

Control inputs to the joints are composed of both feedback PD control and neural network sub-systems components. The outputs of the feedback and neural subsystems are summed to obtain the total control command, τ, representing torques applied at the joints. The feedback controller is based on errors between the desired joint states and actual joint states. The neural network output N is trained to minimize a quadratic cost function, when PD or computed torque controllers are used. Consequently, as the neural network-based sub-system is trained, the feedback contribution to the total control command steadily diminishes and, in the process, the neural network learns the inverse dynamics of the system (in the sense that it can compute the required control command for a desired change in the system output). This type of neural network training paradigm is referred to as feedback-error-learning. The dynamic equation for an n-link rigid robot arm in vectorial form is

$$T = M(q)\ddot{q} + V(q,\dot{q}) + F(\dot{q}) + G(q) = M(q)\ddot{q} + Q(q,\dot{q}) \qquad (2.70)$$

Given a desired trajectory defined in terms of joint variables, namely $(q_d, \dot{q}_d, \ddot{q}_d)$, the next step in our control problem is to compute the necessary torques for the joints so that the manipulator follows the desired trajectory. Equation (2.70) represents the inverse dynamics of a robotic arm, that is, given a set of joint variables, $(q, \dot{q}, \ddot{q})$, we can obtain the corresponding torque values to be used to drive the actuator. The manipulator's direct dynamics can readily be obtained as follows:

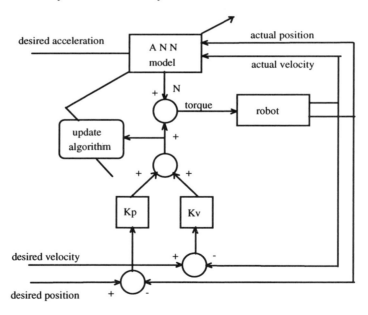

Figure 2.8 Neuro-controller based computed-torque architecture

$$M(q)\ddot{q} = T - Q(q,\dot{q}) \tag{2.71}$$

and then

$$\ddot{q} = M^{-1}(q)T - M^{-1}(q)Q(q,\dot{q}) = R(q,\dot{q},T) \tag{2.72}$$

Equation (2.72) refers to a nonlinear mapping from the robot input (joint torque T) to the robot output (joint motion). The robot inverse dynamics can now be written as

$$T = R^{-1}(q,\dot{q},\ddot{q}) \tag{2.73}$$

where the transformation R^{-1} is a nonlinear mapping from the joint coordinate space to the joint torque space. For an n-link manipulator, this is defined by the following mapping:

$$R^{-1} : \Pi^{3n} \rightarrow \Pi^n \tag{2.74}$$

In practice, robot dynamics cannot be modelled exactly. An estimated model $\hat{R}^{-1}$ is used to predict the feedforward torques and servo-feedback is usually included to bring robustness to the overall control scheme.

The system dynamics are not time invariant and undergo changes such as variations in payloads, changes in the friction coefficients of the joints, etc. Hence, the model estimate $\hat{R}_p^{-1}$ has to be modified accordingly in order to accommodate these changes. To achieve this, an adaptive or a learning control element is usually associated with the control structure. In the novel control architecture developed, $\hat{R}_p^{-1}$ is modelled by artificial neural networks. The nonlinear inverse transformation can be decoupled into n less complex transformations, namely:

$$T = R^{-1}(q,\dot{q},\ddot{q}) = \begin{bmatrix} r_1^{-1}(q,\dot{q},\ddot{q}) \\ \dots \\ \dots \\ \dots \\ r_n^{-1}(q,\dot{q},\ddot{q}) \end{bmatrix} \tag{2.75}$$

where $\hat{r}_i^{-1}(q,\dot{q},\ddot{q}) : i \in [1, n]$ defines the inverse dynamics transformation of the corresponding joint:

$$\hat{r}_i^{-1}(q,\dot{q},\ddot{q}) : R^{-1} : \Pi^{3n} \rightarrow \Pi^n \quad \text{with} \quad i \in [1, n] \tag{2.76}$$

Each r_i^{-1} can be modelled by an ANN such that

$$T = R^{-1}(q,\dot{q},\ddot{q}) = \begin{bmatrix} r_1^{-1}(q,\dot{q},\ddot{q}) \\ ...... \\ ...... \\ ...... \\ r_n^{-1}(q,\dot{q},\ddot{q}) \end{bmatrix} = \begin{bmatrix} N_1(q,\dot{q},\ddot{q},\theta_1) \\ ...... \\ ...... \\ ...... \\ N_n(q,\dot{q},\ddot{q},\theta_n) \end{bmatrix} \qquad (2.77)$$

where $(\hat{\ })$ denotes the estimated models, $N_i(.)$ represents the output of each ANN model used to realize the nonlinear mapping $r_i^{-1}(.)$ and Θ terms denote the set of adjustable weights of the corresponding ANN model.

The algorithm of the neural controller can be summarised as follows:

(i)   Present the desired inputs $q_d(t), \dot{q}_d(t), \ddot{q}_d(t)$.

(ii)   The forward phase of the BP algorithm is executed and the output of the ANN model is a torque N.

(iii) The error vector is added to the torque N and the output in this stage is the torque applied to the robot system.

(iv)   The robot system will produce the actual value of $q_d(t), \dot{q}_d(t), \ddot{q}_d(t)$. Compute the error vector between the desired and the actual vector of the position and velocity.

(v)   The error vector is applied for the computation of the learning (BP, RPE, PRPE, EKF or QR-WRLS) algorithm.

(vi)   The operation is repeated until convergence.

2.7.1 Control law

Using (2.77), a control law is defined as

$$T = \hat{R}^{-1}(q,\dot{q},\ddot{q}) + T_{fd} \qquad (2.78)$$

where

$$T_{fd} = K_p(q_d - q) + K_v(\dot{q}_d - \dot{q}) = K_d E + K_v \dot{E} \qquad (2.79)$$

The computation of $\hat{R}^{-1}(q,\dot{q},\ddot{q})$ requires the measurement of $\ddot{q}$ in addition to q and $\dot{q}$. One approach is to use $\ddot{q}_d$ instead of differentiating the velocity vector $\dot{q}$ using a first order filter. The arm initially starts moving with a PD servo-control. Then, as the data from the sensors are processed to train the inverse model $\hat{R}^{-1}$, feedforward torques gradually start building up and improve the trajectory following performance.

Substituting the control law (2.78) into (2.76), we obtain the system error equation:

$$K_p E + K_v \dot{E} = \tilde{R}^{-1}(q,\dot{q},\ddot{q}) \tag{2.80}$$

where $\tilde{R}^{-1} = R^{-1} - \hat{R}^{-1}$ denotes the errors between the actual inverse dynamics R^{-1} and the estimated model $\hat{R}^{-1}$:

$$\tilde{R}^{-1}(q,\dot{q},\ddot{q}) = \begin{bmatrix} \tilde{r}_1^{-1}(q,\dot{q},\ddot{q},\Theta) \\ \cdots \\ \cdots \\ \tilde{r}_n^{-1}(q,\dot{q},\ddot{q},\Theta) \end{bmatrix} \tag{2.81}$$

where $\tilde{r}_i^{-1}(q,\dot{q},\ddot{q},\Theta)$ denotes the error in inverse dynamic modelling for each joint and Θ is the adaptive weight vector of the corresponding ith ANN model.

Using (2.80) and (2.81), the error dynamics for each joint can be written as follows:

$$\ddot{e} + k_p e + k_v \dot{e} = \tilde{r}_i^{-1}(q,\dot{q},\ddot{q},\Theta) \quad \text{for} \quad i = 1,....,n \tag{2.82}$$

$(\ddot{e},\dot{e},e)$ denote the acceleration, velocity and position errors at joint (i), respectively, k_p and k_v are the individual servo-gains, respectively.

As shown in [26,27,15], the state space representation of (2.82) is given by

$$\dot{x} = Ax + Bk\tilde{r}^{-1}(q,\dot{q},\ddot{q},\Theta) \tag{2.83}$$

$$e = cx \tag{2.84}$$

where $x = [e,\dot{e}]^T$ is the state vector.

For a given symmetric positive definite matrix V, there exists another symmetric positive definite matrix P which satisfies the Liapunov equation

$$A^T P + PA = -V \tag{2.85}$$

and the output relation

$$c = B^T P \tag{2.86}$$

Hence the gradient update law takes the form

$$\dot{p} = -\Gamma \frac{\partial \tilde{r}^{-1}(q,\dot{q},\ddot{q},\Theta)}{\partial p} B^T Px \tag{2.87}$$

$$\dot{p} = -\Gamma \frac{\partial N(q,\dot{q},\ddot{q},\Theta)}{\partial p} B^T Px \tag{2.88}$$

where $N(q,\dot{q},\ddot{q},\Theta)$ is the output of the *ANN* model. Due to the nonlinear parametric dependence of the term $N(q,\dot{q},\ddot{q},\Theta)$, the computation of the partial derivative term in (2.88) requires the use of the so-called backpropagation algorithm. The closed loop adaptive system given by (2.83), (2.84) and (2.88) defines a coupled nonlinear system of differential equations, and this makes a global convergence and stability analysis of the closed loop system difficult. However, through the use of linearization techniques, local properties of the system dynamics can be studied. Subject to smoothness of the nonlinear operators in (2.83), (2.84) and (2.88), a linearized system whose stability properties are identical to the local stability properties of (2.83), (2.84) and (2.88) can be derived as explained below.

2.7.2   Local stability and convergence analysis
In order to linearize the system defined by (2.83), (2.84) and (2.88) around a nominal trajectory, we choose the nominal values $x^* = x_0 = 0$ and $\Theta^* = \Theta_0$. This choice corresponds to a condition where the system operates in the vicinity of the desired trajectories $(q_d,\dot{q}_d,\ddot{q}_d)$ and the ANN model parameters are close to their desired values. The perturbation vector can now be defined as, $\tilde{x} = x_0 - x = -x$, which is actually the tracking error vector, and $\tilde{\omega} = \omega_0 - p$, which is the parameters error vector.

The linearized update equation for the parameter vector w can be written as follows:

$$\tilde{p} = -\Gamma \Psi B^T Px \tag{2.89}$$

where Ψ is considered as the regressor vector of the error dynamics. Based on this linearized model, the local convergence and stability properties of (2.83), (2.84) and (2.88) can be analysed.

2.8   SIMULATION RESULTS

A simple two-degree-of-freedom manipulator model was used to compare the performances of the QR-WRLS algorithms against other ANN controllers using BP and RPE weight estimation. The manipulator was modelled as two rigid links of lengths $l_1 = 0.5\,\text{m}$ and $l_2 = 0.5\,\text{m}$ with point masses $m_1 = 10\,\text{kg}$ and $m_2 = 8\,\text{kg}$ at the distal ends of the links. The pay-load was considered as a point mass m concentrated at the end of the link. The simulation was carried out using a fourth order Runge-Kutta algorithm, with a step size $h = 0.005\,\text{s}$.

The desired joint position trajectory was chosen as ($a_0 = 0.5$, $b_0 = 0.2$):

$$q_{1d} = a_0 + b_0 \sin(t) + b_0 \sin(2t) \qquad \text{for the first joint}$$

$$q_{2d} = a_0 + b_0 (\cos 4t + \cos 6t) \qquad \text{for the second joint}$$

The task for the second joint is to arrive at the final position q_{2f} in exactly four seconds. Figures 2.9 and 2.10 show the desired and actual trajectories for the first and the second joints, respectively. Figures 2.11 and 2.12 show the RMS trajectory errors for joints 1 and 2 respectively. The desired and actual trajectory in Cartesian space is plotted in Figure 2.13 and in three-dimensional form in Figure 2.14.

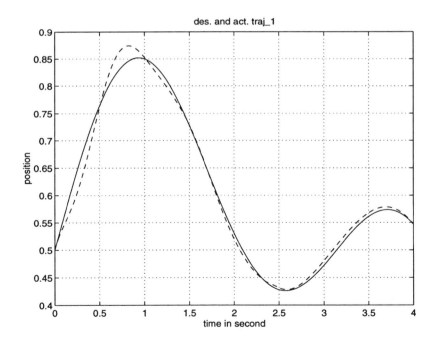

Figure 2.9 Desired and actual trajectory for joint 1

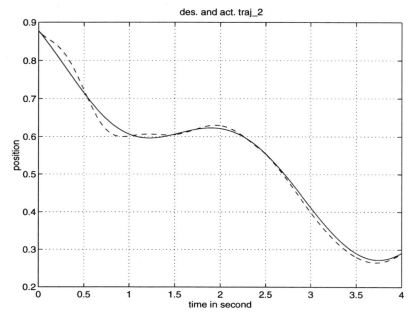

Figure 2.10 Desired and actual trajectory for joint 2

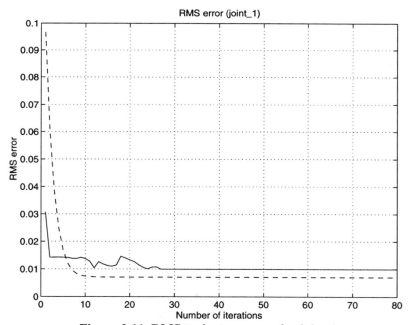

Figure 2.11 RMS trajectory error for joint 1

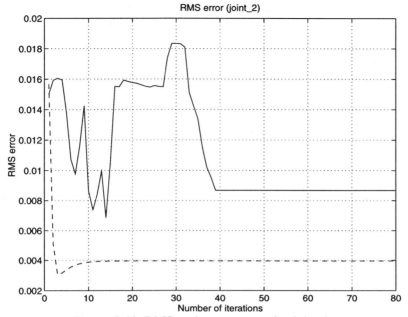

Figure 2.12 RMS trajectory error for joint 2

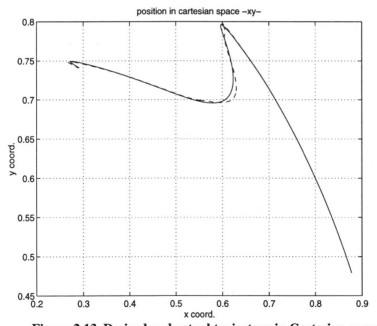

Figure 2.13 Desired and actual trajectory in Cartesian space

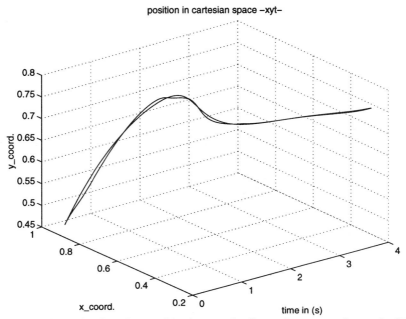

Figure 2.14 Desired and actual trajectory in Cartesian space shown in 3D

A root mean square error method has been used to evaluate the performance of the adaptive controllers based on the alternative ANN models (BP, RPE and QR-WRLS). The RPE algorithm is shown in Figures 2.11 and 2.12 to converge in fewer iterations than the BP algorithm, but it poses a greater computational burden. Furthermore, BP and RPE algorithms suffer from initialization problems. Table 2.1 clearly shows the superior performance of the QR-WRLS algorithms. One hundred sets of initial weights were chosen between [−1, 1]. Some initial weights will lead to a solution, and others will not. This algorithm does not suffer from initialization problems (almost all weights can be used in practice to approximate the ideal initial weights), and it is robust numerically, as no matrix inversion is done.

Table 2.1 Convergence results for various initial weights

			QR-WRLS		
Implementation	BP	RPE	PN-WRLS	PL-WRLS	P Net-WRLS
No. of convergence	20	8	60	96	81
Iteration to convergence	20	4	4	4	10

2.9 CONCLUSION

This work has presented a new approach towards adaptive robot control based on neural network paradigms. The proposed neuro-controller architecture has been applied to the trajectory tracking control of a robot manipulator. The RPE algorithm was used to speed-up the poor convergence with the BP algorithm. However, because both the BP and RPE algorithms suffer from initialization problems, a new algorithm, QR-WRLS was developed which showed several improvements (no initialization problems, numerically stable, can be implemented in parallel, etc.).

The question of weight estimation has been studied in depth. Four weight-update schemes have been considered. These update, respectively, one weight at a time (BP, RPE), all weights connected to one node at a time (PRPE, PN-WRLS), all the weights of the same layer at a time (PL-WRLS) and all the weights of the network simultaneously (PNet-WRLS), for the case of one output.

REFERENCES

[1] S. Dubowsky and D. Desforges (1979), The application of model-referenced adaptive control to robotic manipulators, *J. Dyn. Sys. Meas. and Cont.*, **101**, 193-200.

[2] A. Koivo (1981), Control of robotic manipulator with adaptive controller, *IEEE Conf. on Dec. and Cont.*, San Diego, CA.

[3] T. C. Hsia (1986), Adaptive control of robot manipulators - a review, *IEEE Int. Conf. Robotics and Automation*, San Francisco.

[4] C. H. An, C. G. Atkeson and J. M. Hollerbach (1985), Estimation of inertial parameters of rigid body links of manipulators, *IEEE Conf. Dec. and Cont.*, Fort Lauderdale.

[5] P. Khosla and T. Kanade (1985), Parameter identification of robot dynamics, *IEEE Conf. Dec. and Cont.*, Fort Lauderdale.

[6] J. J. Craig, P. Hsu and S. Sastry (1986), Adaptive control of mechanical manipulators, *IEEE Int. Conf. Robotics and Automation*, San Francisco.

[7] R. H. Middleton and G. C. Goodwin (1986), Adaptive computed-torque control for rigid-link manipulators, *25th IEEE Conf. Dec. and Cont.*, Athens, Greece.

[8] P. Hsu, S. Sastry, M. Bodson and B. Paden, (1987), Adaptive identification and control of manipulators with joint acceleration measurements, *IEEE Conf. Robotics and Automation*, Raleigh, NC.

[9] W. Li and J. J. E. Slotine (1988), Indirect adaptive robot control, *5th IEEE Conf. Robotics and Automation*, Philadelphia.

[10] J. J. E. Slotine and W. Li (1989), Composite adaptive manipulator control, *Automatica*, **25**, 4, 509-519.

[11] J. J. Craig, P. Hsu, and S. Sastry (1987), Adaptive control of mechanical manipulators, *Int. J. Robotics Res.*, **6**, 2, 10-20.

[12] J. J. E. Slotine and W. Li (1987), On the adaptive control of robot manipulators, *Int. J. Robotics Res.*, **6**, 3, 49-59.

[13] R. Kelly and R. Carelly (1988), Unified approach to adaptive control of robotic manipulators, *Proc. IEEE Conf. Dec. and Cont.*, **2**, 1598-1603.

[14] I. Landau and R. Horowitz (1988), Synthesis of adaptive controllers for robot manipulators using a passive feedback systems approach, *IEEE Conf. Rob. and Autom.*, 1028-1033.

[15] R. Ortega and M. Spong (1988), Adaptive motion control of rigid robots: a tutorial, *Proc. IEEE Conf. Dec. and Cont.*, **2**, 1575-1584.

[16] D. E. Rumelhart *et al.* (1986), *Parallel distributed processing*, MIT Press, Cambridge MA.

[17] M. Kawato *et al.* (1988), Hierarchical neural network model for voluntary movement with application to robotics, *IEEE Contr. Syst. Mag.*, 8-16.

[18] M. Kawato, K. Furukawa and R. Suzuki (1987), A hierarchical neural network model for control and learning of voluntary movement, *Biol. Cybern.*, **57**, 169-185.

[19] D. Psaltis, A. Sideris and A. Yamamura (1988), A multilayer neural network controller, *IEEE Contr. Syst. Mag.*, 17-21.

[20] J. J. E. Slotine and W. Li (1988), Adaptive manipulator control: A case study, *IEEE Trans. Automat. Contr.*, **33**, 11, 995-1003.

[21] A. M. S. Zalzala and A. S. Morris (1989), A neural network approach to adaptive robot control, *Int. J. Neural Networks*, **2**, 17-35.

[22] A. Y. H. Zomaya, M. E. Suddaby and A. S. Morris (1992), Direct neuro-adaptive control of robot manipulators, *Proc. IEEE Int. Conf. on Robotics and Automation,* Nice, France, May 1992, 1902-1907.

[23] S. Khemaissia and A. S. Morris (1993), Neuro-adaptive control of robotic manipulators, *Robotica*, **11**, 465-473.

[24] G. C. Goodwin and K. S. Sin (1984), *Adaptive filtering prediction and control*, Prentice Hall, Englewood Cliffs, NJ.

[25] K. J. Astrom and B. Wittenmark (1989), *Adaptive control*, Addison Wesley, Reading, MA.

[26] K. C. Funahashi (1989), On the approximate realization of continuous mappings by neural networks, *Neural Networks*, **2**, 183-192.

[27] M. S. K. Hornik and H. White (1989), Multilayer feedforward networks are universal approximators, *Neural Networks*, **2**, 359-366.

[28] J. Makhoul, R. Schwartz and A. El-Jaroudi (1989), Classification capabilities of two-layer neural nets, *Proc. IEEE Int. Conf. on Acoustics, Speech and Signal Proc.*, Glasgow, 635-638.

[29] S. Haykin (1991), *Adaptive filter theory*, Prentice Hall.

[30] L. Ljung and T. Soderstrom (1983), *Theory and practice of recursive identification*, MIT Press, Cambridge, MA.

[31] S. A. Billings, H. B. Jamalludin and S. Chen (1991), A comparison of the backpropagation and recursive prediction error algorithms for training neural networks, *Mech. Syst. and Signal Proc.*, **5**, 3, 233-255.

[32] S. Chen, S. A. Billings and P. M. Grant (1991), Nonlinear system identification using neural networks, *Int. J. Control*, **51**, 6, 1191-1214.

[33] G. Golub and C. van Loan (1983), *Matrix computations*, Johns Hopkins University Press, Baltimore, MD.

[34] S. S. Sastary and A. Isidori (1989), Adaptive control of linearizable systems, *IEEE Trans. on Autom. Control*, **34**, 1123-1131.

[35] G. Feng, M. Palaniswami and Y. Yao (1991), Adaptive tracking control of robot manipulators in cartesian space coordinates, *Proc. of the American Control Conf.*, **33**, 932-937.

3

Neural servo controller for position, force and stabbing control of robotic manipulators

T. Fukuda[1], T. Kurihara[2], T. Shibata[3], M. Tokita[4] and T. Mitsuoka[4]
[1] Department of Mechano-Informatics and Systems, Nagoya University, Japan
[2] IBM Japan
[3] Mechanical Engineering Laboratory, MITI, Japan
[4] National Kisarazu Technical College, Japan

3.1 INTRODUCTION

The central nervous system of a human being can generate appropriate actions by the parallel processing of various perceptual information [1, 2]. It also has advanced learning and flexible adaptation capabilities under unknown environments. The authors have previously studied a pushing-motion control of a robotic manipulator with one degree of freedom using a neural network model [3]. They showed that it is necessary to consider the force control of the robotic manipulator in more complicated situations; for example, a micro robotic manipulator in biological cell manipulation drills and breaks biological cell membranes without destroying them, with dynamics which are completely different from the normal cases of so-called stabbing phenomena. Under these conditions, it is difficult to model such a nonlinear process by a clear mathematical method and make the stabbing control handling the reactive force control smoother in the transient period before and after the breakage. Instead, a hybrid position/force control system is needed: here an integrated process combines the force and position information appropriately [4]. In the event that the objects are unknown, a system with fixed gains has limited applicability for these adaptations.

This chapter presents a neural network-based robotic motion controller, the neural servo controller. References [5, 6] show the effectiveness of a neural servo controller for hybrid position/force control and stabbing control of the robotic manipulator. This chapter deals with stabbing control. The manipulator has a needle at its tip and breaks a membrane/thin film that has nonlinear characteristics before and after the breaks. Conventional control methods could not deal with nonlinear control problems like this, since the dynamics of the objects change drastically [7,8]. Therefore, the goal of this chapter is to control nonlinear systems with uncertainties by employing the self-organization ability of the neural network. The neural servo controller is applied for the self-organizing system of the sensor fusion of both force and position information, so as to perceive whether the membrane/thin film is broken or not and to switch the control mode from the position-control to the force-control mode. The neural servo controller does not need a strict mathematical model of the robot, since it has the capability to learn the model.

In applying neural networks to robotic problems, we have to take account of stability problems. While sensing and perception capabilities have not involved the dynamic characteristics so far, the dynamic process with feedback loops must have stability problems. At present, stability problems are still an open question. Moreover, optimality is sometimes required, but the neural network approach seeks feasibility first and then better performance to achieve optimality, which is more difficult to solve by conventional control strategies.

Section 3.2 describes mathematical models of the robotic manipulator and the membrane/thin film that are used for experiments. The models are used for simulations. Section 3.3 explains the neural network model used for the neural servo controller. A four-layered neural network is used. An auto-regressive form of the error back-propagation method is applied to train the neural network. In Section 3.4, simulations of image pattern recognition and stabbing control are performed to show the effectiveness of the neural network. In Section 3.5, experiments on stabbing control are described which demonstrate the effectiveness of the neural servo controller. The last section concludes this chapter.

3.2 MATHEMATICAL MODEL

In this study, the concept of a neural servo control system is employed in order to perform hybrid control, including the position and force control shown in Figure 3.1. The model of a manipulator with one degree of freedom interacting with an object and the mass effect is shown in Figure 3.2. It is assumed that the object can be represented by a linear model with damping coefficient $C*$ and spring constant $K*$. After breakage of the object by stabbing control, these damping and spring constants will be assumed to be 0. The state equation and the output equation can be expressed, employing the dynamic characteristics of the actuator and the load, as follows:

Amplifier:

$$V(t) = A_m * u(t) \tag{3.1}$$

DC motor:

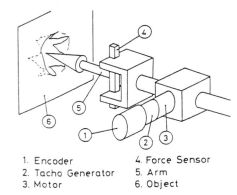

1. Encoder   4. Force Sensor
2. Tacho Generator   5. Arm
3. Motor   6. Object

Figure 3.1 Illustration of a robotic manipulator with one degree of freedom

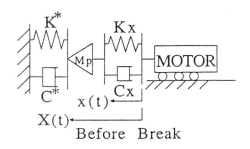

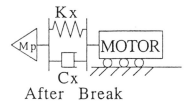

Figure 3.2 Equivalent model of the robotic manipulator

$$L_a * \mathrm{d}\, i(t)/\mathrm{d}\, t + R_a * i(t) + E(t) = V(t) \tag{3.2}$$

Counter-electromotive force:

$$E(t) = K_e * \theta_m(t) \tag{3.3}$$

Balance in DC motor:

$$J_m * \ddot{\theta}_m(t) + B_m * \dot{\theta}_m(t) = T(t) - T_l(t) \qquad (3.4)$$

Torque:

$$T(t) = K_t * i(t) \qquad (3.5)$$

Balance in load:

Mass M:

$$M * \ddot{x}(t) + B_l * \dot{x}(t) + C_x * \left\{ \dot{x}(t) - \dot{X}(t) \right\}$$
$$+ k_x * \left\{ x(t) - X(t) \right\} = n/r * T_l(t) \qquad (3.6)$$

Mass M_p:

$$M_p * \ddot{X}(t) + C^* * \dot{X}(t) + K^* * X(t)$$
$$- C_x * \left\{ \dot{x}(t) - \dot{X}(t) \right\} - K_x * \left\{ x(t) - X(t) \right\} = 0 \qquad (3.7)$$

Relationship with turning angle of the motor and movement of the manipulator:

$$\theta_m(t) = n/r * x(t) \qquad (3.8)$$

where
A_m: gain of amplifier
B_m: frictional resistance of the load including the reduction gear
M: mass of the load
M_p: mass of the top of the manipulator
B_l: frictional resistance of the load
R_a: armature resistance of the motor
K_e: constant of the back electromotive force
K_t: torque constant of the motors
L_a: inductance of the motor
K^*: spring constant of the object
C^*: damping coefficient of the object
K_x: spring constant of the force sensor and the manipulator
C_x: damping coefficient of the force sensor and the manipulator
r: radius of the pinion gear
n: gear reduction ratio
$U(t)$: input voltage
$T_1(t)$: load torque
$T(t)$: generated torque
$E(t)$: back electromotive force

$v(t)$: input to the motor
$x(t)$: position of the manipulator
$X(t)$: bend of the force sensor
$i(t)$: electric current of the motor
m: turning angle of the motor

The dynamic equations are as follows:

$$\ddot{x}(t) = -a_1 x(t) - a_2 \dot{x}(t) - a_3 X(t) - a_4 \dot{X}(t) + b_1 U(t) \tag{3.9}$$

$$\ddot{X}(t) = -a_5 x(t) - a_6 \dot{x}(t) - a_7 X(t) - a_8 \dot{X}(t) \tag{3.10}$$

where

$$a_0 = n^2 * J_m + r^2 * M$$

$$a_1 = r^2 * k_x / a_0$$

$$a_2 = \left(r^2 * R_a * (B_l + C_x) + n^2 * k_e * k_t \right) \Big/ (R_a * a_0)$$

$$a_3 = -r^2 * k_x / a_0$$

$$a_4 = -r^2 * C_x / a_0$$

$$a_5 = k_x / M_p$$

$$a_6 = C_x / M_p$$

$$a_7 = \left(k_x + K^* \right) \Big/ M_p$$

$$a_8 = \left(C_x + C^* \right) \Big/ M_p$$

and

$$b_1 = A_m * n * r * k_t \Big/ (R_a a_0)$$

The state is given by

$$\dot{X}(t) = AX(t) + Bu(t) \tag{3.11}$$

where

$$X = \begin{bmatrix} x(t) & \dot{x}(t) & X(t) & \dot{X}(t) \end{bmatrix}^T$$

$$A = \begin{bmatrix} 0 & 1 & 0 & 0 \\ -a_1 & -a_2 & -a_3 & -a_4 \\ 0 & 0 & 0 & 1 \\ -a_5 & -a_6 & -a_7 & -a_8 \end{bmatrix}$$

and

$$B = \begin{bmatrix} 0 \\ b_1 \\ 0 \\ 0 \end{bmatrix}$$

The output equation is expressed by using the hybrid ratio S_1 between the position and force control as follows:

$$y(t) = s_1 x(t) + (1 - s_1) F(t) \qquad (0 \le s_1 \le 1) \qquad (3.12)$$

and

$$F(t) = C_x \{\dot{x}(t) - \dot{X}(t)\} + k_x \{x(t) - X(t)\} \qquad (3.13)$$

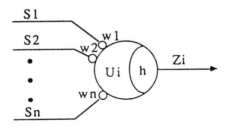

Figure 3.3 Neuron model

3.3 NEURAL NETWORK MODEL

3.3.1 Neuron model

Figure 3.3 shows a neuron model that is a primary element of a multi-layered network. The potential of the neuron is given by the weighted summation of all the former neuron units. A unit has a sigmoid function, and the output value sets the limits $-1 < f(x) < 1$

continuously. However, a unit in the input layer is characterized by taking a linear output function, expressed as follows:

$$u_i(t) = \sum w_{ij}(t) S_j(t) \tag{3.14}$$

$$z_i(t) = f\{u_i(t)\} \tag{3.15}$$

and

$$f(x) = \{1 - \exp(-x)\} / \{1 + \exp(-x)\} \tag{3.16}$$

where
$U_i(t)$: the potential of unit i
$Z_i(t)$: the output value of unit i
$W_{ij}(t)$: the weight on connection from unit j to unit i
$S_j(t)$: the input signal
$f(x)$: the output function

3.3.2 Network structure

Figure 3.4 shows a multi-layered neural network model. This model has four layers and feeds the forward structure in one direction from the input layer to the output layer, such that it forbids connection within a layer or connections backwards between layers [9].

3.3.3 Learning procedure

In this chapter, we adopt an auto-regressive type of backpropagation method [1,10] to adjust the connection weights that realize the optimized network connection system described previously. The total error that is to be minimized, E, is defined as

$$E = 1/2 \sum (z_i - d_i)^2 \tag{3.17}$$

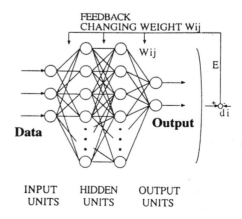

Figure 3.4 A four-layered neural network model

where Z is an actual value of output unit i, and d is its desirable output. A weight value $w_{ij}(k)$ will change gradually with $w_{ij}(k+1)$ by one learning process. $\Delta w_{ij}(k+1)$ is defined as

$$\Delta W_{ij}(k+1) = -\eta\, \partial E\, /\, \partial W_{ij} + \alpha \Delta W_{ij}(k) + \beta \Delta W_{ij}(k-1) + \gamma \Delta W_{ij}(k-2) \qquad (3.18)$$

where η, α, β and γ are positive or negative constants. By tuning the coefficients η, α, β and γ appropriately, the network learning capability and adaptability will be significantly improved. If $\beta = 0$ and $\gamma = 0$, the method becomes a form of the conventional error-backpropagation method.

3.4 SIMULATION

3.4.1 Recognition capability of neural network model
In order to assess the discriminating capabilities of the four-layered neural network model and the learning coefficients (η, α, β and γ) of the auto-regressive form of the back-propagation method that affect error convergence, we have carried out extensive computer simulation for pattern recognition. The neural network model contains eighteen input units, three output units and one to twenty units within the two hidden layers. We employed three possible image-processed data as input, so that three output units gave three classified patterns. The task is specified by determining the desired state of the output units. The neural network repeatedly learns to recognize the three image-processed data groups by firing the specific output unit corresponding to the desired output state.

3.4.2 Stabbing control
The neural network model applied to stabbing control contains three input units, which are the desirable force, position and their hybrid inputs, seven units within each of two hidden layers and two output units connected to the robotic manipulator. One of the output signals drives the plant, while the other gives the force and position hybrid ratio S_1. Therefore, in this case, the neural network can act as a feed forward controller and, simultaneously, as a perception system. Figure 3.5 shows a proposed neural servo control system. The hybrid ratio S_1 gives the breakage and the nonbreakage states of the object by the manipulator: if the robotic manipulator touches the target, the desirable hybrid ratio follows $S_1=1.0$. After the manipulator breaks the target, the ratio is then $S_1=0.0$. If the neural network architecture is successfully trained, then the robot controller can recognize the break for itself.

The parameters of the neural network are employed after the learning process through the step response and are fixed during stabbing control.

3.4.3 Simulation results
Figures 3.6, 3.7 and 3.8 show the simulation results of the image-pattern recognition. Figure 3.7 shows that the actual output values have converged to desirable

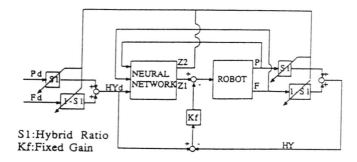

Figure 3.5 The neural servo control system

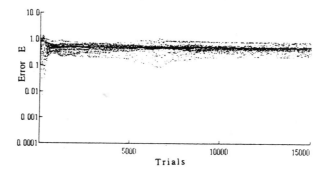

Figure 3.6 A simulation result of pattern recognition
($\eta = 0.25$, $\alpha = 1.2$, $\beta = -1.0$, $\gamma = 0$, ten hidden units)

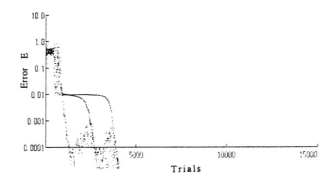

Figure 3.7 A simulation result of pattern recognition
($\eta = 0.8$, $\alpha = 1.2$, $\beta = -0.4$, $\gamma = 0$, ten hidden units)

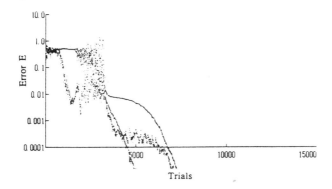

Figure 3.8 A simulation result of pattern recognition
($\eta = 1.51$, $\alpha = 0.5$, $\beta = -0.2$, $\gamma = -0.4$, three hidden units)

output values, while Figure 3.6 shows that the squared error does not decrease because unsuitable learning coefficients were used. The choice of the learning method chosen may greatly affect the performance of the control system in terms of adaptation, learning capabilities and convergence. Unsuitable coefficients in the backpropagation method lead the neural network architecture to perform badly, as shown by the example given above. Using the fourth term, the auto-regressive form of the backpropagation method provides one way of solving the problem of stagnation of the learning process and avoiding local minima. Figure 3.8 shows a simulation result with the fourth term of the auto-regressive form of the backpropagation method. In this case, we need only three hidden units to obtain a good result from the trained neural network, while we must have at least eight hidden units for good performance without the fourth term.

Figures 3.9 and 3.10 provide examples of stabbing control simulation. The simulations have been carried out for the spring stiffness of the objects, $K^* = 50$ N/m and $K^* = 10$ N/m. Figure 3.9 shows the simulation results for stabbing control by the fixed feedback gain. Figure 3.9(c) is one example of the failure of stabbing control. Figure 3.10 shows the results obtained under the same conditions by using the proposed neural servo control system. The neural network outputs, unit 1 and unit 2, show the input to the motor and the hybrid position/force ratio S_1, respectively. In order to recognize the breakage and the maximum displacement of the force output, it is necessary to consider breaking conditions of objects and to set the control system so that it can recognize breakage when the neural servo control system is not used. In this case, it might be possible not to detect the breakage shown in Figure 3.9 (c). This diagram shows that the force still remains after breakage by itself. The control with fixed gains gives a better performance of force to breakage at $K^* = 50$ N/m, but the neural network control system is better overall for reaction control in conjunction with breakage recognition.

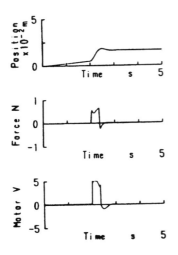

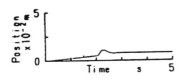

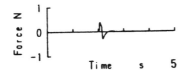

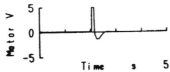

(a) K^*=50 N/m

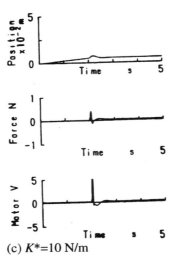

(c) K^*=10 N/m

Figure 3.9 Simulation results of stabbing control by fixed feedback gains

(b) K^*=10 N/m

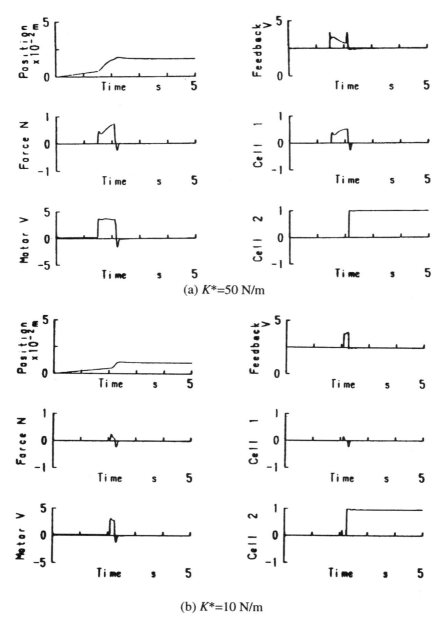

(a) K^*=50 N/m

(b) K^*=10 N/m

Figure 3.10 Simulation of stabbing control using the neural servo controller

Figure 3.11 The experimental equipment of the neural servo control system

3.5 EXPERIMENTAL RESULTS

3.5.1 Experimental method

Figure 3.11 shows the experimental equipment for the neural servo control system, consisting of the DC motor as an actuator, a sharp 3 mm-diameter needle made from brass and a drive mechanism with one degree of freedom by the rack and pinion gear structure. The manipulator can move 150 mm with maximal length. The output of the position and the thrust force are measured by the encoder and the strain gauge, respectively, attached at the root of the needle. The manipulator can be controlled by the microcomputer.

A vinyl film and a rubber balloon are selected as unknown objects. The tip of the needle starts a distance away from the object, and only velocity control of the manipulator is carried out until the force sensor detects a prescribed value of the force. Then stabbing control is applied. The neural network adjusted by the simulation model is used for the experiments.

3.5.2 Experimental results

Employing the experimental equipment of the manipulator shown in Figure 3.11, extensive experiments have been carried out for stabbing control. Some of the exp erimental results are shown in Figures 3.12 and 3.13, in which the broken line represents the recognized point of breakage, while the solid line represents the output of the system. In these experiments, the neural servo control system could provide stable control of the stabbing motion against both the vinyl film and the rubber balloon, and recognize the stabbing processes. However, the fixed gain control system, as expected, gives similar results in simulations where the manipulator stopped before breakage, as shown in Figure 3.12(b). Thus, the neural network-based control system gives a stable control performance and better recognition of breakage, even where objects vary for stabbing control.

3.6 CONCLUSIONS

In this chapter, we have proposed a neural servo controller for a robotic manipulator which uses the self-organizing capabilities of neural networks, and have shown the adaptive control capabilities for both positioning and force control of the robotic manipulator by the proposed method. Using the proposed neural servo controller, we have demonstrated comparatively stable control for different types of object and sensing of breakage by integrating position and force information through simulation and experimental results. Finally, it is impossible at this moment to prove the stability of the model from the general framework, but a future challenge will be to prove the stability of the simplified system model.

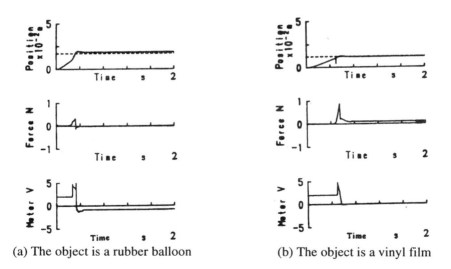

(a) The object is a rubber balloon   (b) The object is a vinyl film

Figure 3.12 Experimental results of stabbing control by fixed feedback gains

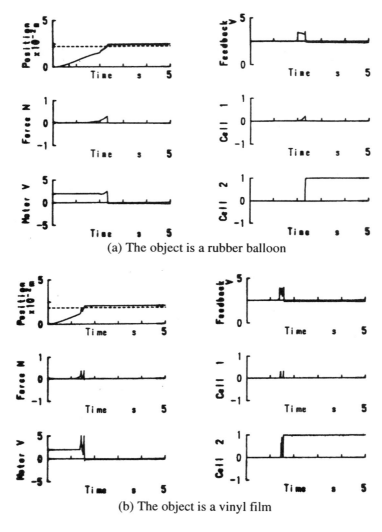

(a) The object is a rubber balloon

(b) The object is a vinyl film

Figure 3.13 Experimental results of stabbing control using the neural servo controller

REFERENCES

[1] D. E. Rumelhart, G. E. Hinton and R. J. Williams, Learning by error, *Nature*, **323**, 533-536 (1986)

[2] T. Fukuda and T. Shibata, Theory and applications of neural networks for industrial control applications, *IEEE Trans. on Industrial Electronics,* **39**(6), 472-489 (1992)

[3] M. Tokita, T. Fukuda, T. Mitsuoka and T. Kurihara, Force control of robotic manipulator by neural network, *Journal of Robotic Society of Japan,* **7**(1), 47-51 (1989)

[4] T. Fukuda, N. Kitamura and K. Tanie, Adaptive force control of manipulators with consideration of object dynamics, *Proc. of IEEE Conf. on Robotics and Automation*, 1543-1548 (1987)

[5] T. Fukuda, T. Kurihara, T. Shibata, M. Tokita and T. Mitsuoka, Application of neural-network-based controller to position, force and stabbing control by robotic manipulator, *JSME Int. Journal, Series III*, **34**(2), 303-309 (1991)

[6] T. Fukuda, T. Shibata, M. Tokita and T. Mitsuoka, Neuromorphic control for robotic manipulators: adaptation and learning, *IEEE Trans. on Industrial Electronics*, **39**(6), 497-503 (1992)

[7] M. Kawato, Y. Uno, M. Isobe, and R. Suzuki, A hierarchical neural-network model for voluntary movement with application to robots, *IEEE Control Systems Magazine*, **8**(2), 8-16 (1988)

[8] P. J. Werbos, Neurocontrol and related techniques, *Hand Book of Neural Computing Applications*, Academic Press, Inc., 345-380 (1990)

[9] K. Funahashi, On the approximate realization of continuous mapping by neural networks, *Neural Networks*, **2**, 183-192 (1989)

[10] N. Watanabe, S. Nagata and K. Asakawa, Fujitsu Laboratories mobile robot control by neural networks and their associated learning algorithm, *Proc. of IECON'88,* 1054-1060 (1988)

4

Model-based adaptive neural structures for robotic control

Ali M. S. Zalzala
Robotics Research Group, Department of Automatic Control and Systems Engineering, University of Sheffield, Mappin Street, P.O. Box 600, Sheffield S1 4DU, United Kingdom

4.1 INTRODUCTION

Real-time robot control has always presented researchers with great difficulties in terms of both the accuracy of the command actions required and also the efficiency by which the commands are obtained. A very important characteristic of the new generation of robotic systems is the presence of intelligent capabilities which is being rapidly supported by fast computing power and adequate sensory equipment.

Conventional controllers for industrial robots are based on independent joint control schemes in which each joint is controlled separately by a simple servo-loop with pre-defined constant gains. Although this control scheme is adequate for simple pick-and-place tasks where only point to point motion is of concern, in tasks where precise tracking of fast trajectories under different payloads is required, the existing control schemes are severely inadequate. When producing the control commands, the coupling between the manipulator joints and the non-linear nature of the system should be taken into consideration. Therefore, in real-time application, ignoring all or parts of the robot non-linear dynamics, or even errors in the parameters of the manipulator, may cause serious deviations and thus render such control strategies inefficient. None the less, although a complicated control scheme such as the computed torque technique could be tolerated through its implementation on fast state-of-the-art computing structures, such constant gains could produce substantial tracking deviations. One solution has been proposed by the introduction of different schemes for adaptive robot control [1]. Nevertheless, the proposed theory has been considered to present too heavy a mathematical burden for execution within the short control cycle usually required for fast trajectory tracking.

The application of the theory of artificial neural networks has emerged in recent years as a very attractive area of research. For the engineering community, the theory of cognition offers a possible solution for different inherent problems for which the massive parallelism and learning abilities associated with neural networks are very much appreciated. Different approaches are reported in the literature for implementing neural-based control algorithms [2,3], including simple PID controllers [4] and the more sophisticated model-reference adaptive controller [5]. In addition, other applications in system identification [6] and non-linear system theory [7] are also reported. Many successful solutions are delivered in the field of pattern recognition [8] where different network designs are available. However, where real-time dynamic systems are concerned, less success has been achieved [9].

The fast and accurate control of articulated robot manipulators forms a difficult challenge to engineering researchers due to the very complicated nature of the overall dynamic system. Problems in the control of robots arise from the vast computational complexities associated with its mathematical formulation, in addition to the need for appropriate adaptive control methods to achieve the required precision and speed [10]. Therefore, the robotics community has been interested in connectionism as one possible solution, where the massive parallelism within the network can reduce the computational burden. In addition the learning abilities make such an approach attractive in the sense of providing compensation for any errors in executing the robot motion. Thus, by producing a robot control model based on the structure of the human brain, it is hoped to inherit the latter abilities and achieve human-like behaviour [11].

Thus, it is recognised that the non-linearities in the robot model make the adaptation of a large network extremely difficult. These difficulties would increase with the arm having more degrees of freedom, and it has always been the case for simulation results to be presented mostly for simple two-joint arms. In the kinematics control of a robot arm, a feed-forward layered network was proposed for the motion tracking of a two-joint arm and later implemented on an actual robot manipulator [12,13]. In addition, a scheme has been proposed for the solution of the inverse kinematics algorithm for a three-joint planar manipulator [14]. Furthermore, artificial neural nets have been proposed for the representation of navigation maps for autonomous path planning [15]. Adaptive dynamic control has been tackled by several researchers as well. A preliminary analysis of the problem has been assessed [16], while different approaches to the solution can be found [3,17].

None the less, in proposing a solution for any neural-based control problem, the difficulty remains twofold. First, a suitable structure must be chosen for the neural network to ensure that no extra complexity is introduced. Second, the system must be able to generalise its operation to include any new situation with a minimum of extra learning processes. This latter is of particular importance in intelligent sensory-based robotic applications where the system is expected to operate independently and in real-time. In such cases, new circumstances, in terms of environmental and operational changes, require the controller to adapt rapidly to perhaps completely new rules. Indeed, any neural structure must comply with such on-line operation if it is to be considered effective.

In this chapter, a new computational model is presented for the learning control of robot manipulators. Certain concepts of the neural network approach have been extracted to provide for a fast real-time controller. The neural controller is designed on a multi-layered network, for which the adaptation for environmental changes could be accommodated via the back-propagation of errors throughout different layers. A symbolic representation of the dynamic equations of motion is considered, for which the proposed controller is shown to be adequate to accommodate for the presence of uncertainties in the model used, as well as any changes in the robot model during real-time applications. In addition, the use of distributed parallel processing concepts greatly simplifies the computations required within each control cycle. Simulation results are reported for the full model of the Unimation PUMA 560 manipulator with six degrees of freedom.

4.2 THE THEORY OF COGNITION

Connectionism offers the opportunity of comparing the operation of engineering systems to the actual behaviour of living beings, thus providing a unique step towards effective artificial intelligence systems. However, one main aim of ongoing research is to recast the control problem in a suitable mathematical form, and, further, to set up a model simulating the nervous system.

A comprehensive survey of different neural networks used in engineering applications can be found in Chapter 1 and elsewhere [18,19], while the following serves only as an introduction to the development of the work reported in this chapter.

4.2.1 Main structure

The main block in a neural network is the neurone node, which is expected to include a mere simple computational task. As shown in Figure 4.1(a), the node is a multiple-input single-output device with the following parameters :

- Inputs to the node: denoted by I_j, $j=1,2,...,n$, which could be external inputs or alternatively outputs of other neurones.
- Weight of each input: denoted by W_{ij}, linking the neurone i with others $j=1,2,....$ These are adapted during use in order to improve performance.
- Combining function (F_c^i): Relates all inputs to the node, and can be either linear or non-linear.
- Activation function (F_a^i)the output of the neurone, which could be connected to any other node in the network, or even connected as a feed-back to the same node.

A one-to-one correspondence of this structure with its physical origin in brain cells can be established [20]. However, it is very interesting to note that although modern high-speed VLSI structures can manipulate data much faster than the human brain, their capabilities as far as learning is concerned is quite restricted. This may be due to the brain operating in a highly parallel fashion, where it has a thousandfold greater connectivity than currently available supercomputers.

A neural network is usually constructed of an input layer of neurones, an output layer and several intermediate, or hidden, layers. The number of hidden layers used, along with the number of neurones in each of them, depends on the learning requirements. In addition, a network may be either fully or sparsely connected. If fully connected, the output of each node in a layer is connected as an input to all nodes in the next layer. Alternatively, a network could be sparsely connected, where an example of a four-layer network is shown in Figure 4.1(b).

Although a general assumption is made for a multi-layered network so that the nodes of each hidden layer are accessed by the nodes of each predecessor layer only, it has been noted that there is no reason why some units may not receive inputs from other units in earlier layers [19].

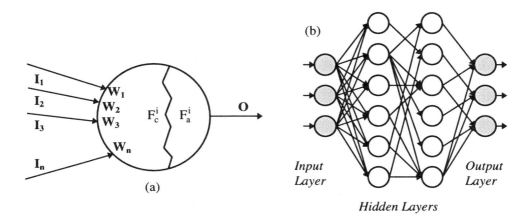

Figure 4.1 (a) A neurone. (b) A multi-layered neural structure

4.2.2 The back-propagation algorithm

Although several different schemes are available for the construction of multi-layered neural nets, the back-propagation algorithm is the most common, and has been reported by many researchers to perform well. The back-propagation learning rule is based mainly on a combination of the least-mean-square and the gradient descent methods. It involves two phases in its operation, as follows :

- Forward phase: includes inputting the desired values, computing throughout the network's layers and producing its outputs.
- Backward phase: where the actual outputs of the network are compared to the desired values, and the errors resulting are propagated back through each of the layers, adjusting the weights of each node.

This process is continued for other inputs until the network learns how to behave in the desired manner. This general form of the back-propagation algorithm is for a fully connected network, while any sparsely connected structure (such as the one shown in Figure 4.1(b)) requires suitable modification.

4.3 DRAWBACKS IN CONTROL APPLICATIONS

The general form of the multi-layered neural network uses the neurone as its basic block.

The fact that the biological origin on which such a neural model is based is still being investigated by neurobiologists is of no real concern to engineering researchers. As far as control systems are concerned, certain concepts of neural networks are readily adapted by engineers as presented to them, and attempts are made to use the rather ambiguous model to solve a required problem. However, such an approach introduces enormous application difficulties as discussed below.

4.3.1 Constructing the network

A typical neural network model consists of an input layer, an output layer and a number of hidden layers of neurones. As illustrated in Figure (4.2), the hidden part of the network can be considered as a black box for which a suitable design must be introduced, preferably to imitate learning capabilities. Therefore, to apply the required engineering theory, certain assumptions are made, including setting up some number of hidden layers, each containing a number of computational neurones. In addition, learning parameters must be given which depend on the learning algorithm used by the network. Once the network design is finished, execution commences by applying the desired inputs and further checking the actual output against the desired values, where the network is expected to learn from the errors. Hence, this execution procedure is repeated iteratively for hundreds or even thousands of times, during which it is hoped that convergence may occur.

If the network design used is appropriate for the type of problem tackled, satisfactory results may be obtained. This is not always the case, however. Although the general assumption of associating the problem complexity with the size of the network used is taken as granted, recent research results indicates a contradiction [21], where simple regular networks have proved extremely difficult to train. Hence, it is concluded that an intelligent guess at the network design and the learning parameters is a *must* to produce good results.

Referring to Figure 4.2, the only well-defined concept is the applied engineering theory, which is very reliable and, indeed, has stood the test of time. What engineers are trying to do, then, is to replace their own well-established theory with an assumed model taken from a not so well-defined biological basis, and further going about executing the new design iteratively which may take many days on the most efficient supercomputer. Nevertheless, if the initial assumption is wrong, the network will never converge and no satisfactory results will be obtained.

4.3.2 The learning algorithm

As indicated in Section 4.2.3, the most popular learning algorithm used in neural applications is the back-propagation of errors [22]. The back-propagation algorithm is reported to give good results in many applications, in particular when classification and pattern recognition is tackled. However, problems still occur when applying this learning procedure, namely, the initial set-up of weights, learning factors and momentum factors, which become more problematic with increasing size of the network. In addition, reasonable doubts are casted on the ability of the gradient descent minimisation procedure to detect global minimas, where such a problem becomes more obvious with the increased dimensionality of the system, thus making the system more prone to be caught in a local minima rather than the required global state. In fact, the gradient descent method is a mathematical procedure with no intelligent capabilities at all to indicate its convergence in the wrong direction [23].

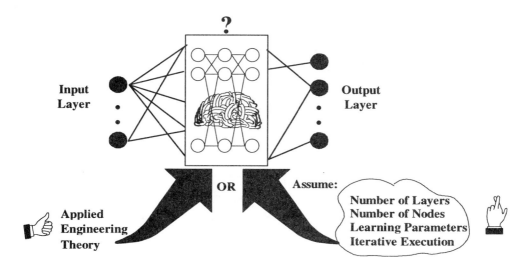

Figure 4.2 The general approach to the application of neural networks

When this learning procedure is used, it is suggested that the neural network will acquire the exact model of the controlled system and then use it to estimate and adapt for any changes in the model, thus effectively controlling the system. However, it seems inappropriate to disregard the well-established, though imperfect, mathematical models available for a robotic manipulator in the hope that a neural structure may learn a perfect form of the model.

4.4 A NEURAL CONTROLLER PROSPECT

The general structure of the neural controller is shown in Figure 4.3, where I and O are the inputs to the network and the outputs from the network, respectively, and subscripts a and d denote the *actual* and *desired* parameters, respectively. Thus, using the desired input parameters, the neural controller produces a set of control commands moving the robot arm to a certain actual position. A feedback of this actual position is used by the arm model (has no learning abilities) producing a set of actual control commands. Finally, the difference between the actual and desired control commands is used to adjust the neural network by propagating the errors back throughout its layers.

The adaptation of the neural controller depends upon the sensory feedback information in each control cycle. Hence, if the model of the actual robot varies from that predicted by the arm model, the network weights are adjusted so as to cancel such a difference. Thus, as time progresses, the controller should correct any errors occurring during tracking of the desired motion.

It should be emphasised that the learning algorithm is applied using the output tracking errors rather than the input errors. Although the latter approach has usually been suggested in the literature, using the difference between the desired robot command-values (O_d) and the actual values (O_a) is seen as more realistic for a practical implementation. Nevertheless, to provide for the errors in the output values as an input to the second phase of the back-propagation rule, the approach for the simulation was to assign an arm computations module (denoted as the arm model in Figure 4.3) to run simultaneously with the controller.

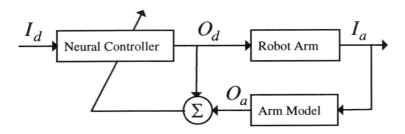

Figure 4.3 The structure of the neural controller

In addition, the neural network is given more intelligent abilities by *injecting* some knowledge of the controlled system into the initial design and operation of the network. This is made possible by replacing the combining function within each neurone, which is usually a summation of inputs multiplied by weights. The new combining function is a non-linear function which is a part of the manipulator mathematical model. However, the activation function remains a continuous sigmoid. Consequently, the computational burden associated with each neurone will depend on the complexity of the combining function derived from the system model.

Since the combining functions are different for each neurone, and each requires a different set of inputs, the network will have sparse connections. In addition, some neurones in a hidden layer may have connections not only from the immediate previous layer but from other previous layers as well.

Taking all the above into consideration, a distributed learning system can be organised where all neurones and every connection is accounted for according to the original system model. Thus, this approach tolerates no ambiguities when designing the network.

4.4.1 The modified back-propagation learning rule

To implement the proposed algorithm, the back-propagation rule must run on several hidden layers, with nodes that are sparsely connected, according to the needs of the mathematical formulation. Thus, the following is assumed:

[1] Initial weights for all connections coming out of all nodes in the network are set to

unity. Since the node connections are sparse and not complete between the hidden layers, setting the initial weights of the network links to unity does not introduce the problem of symmetry. This arrangement has the effect of computing the equations of the robot as a conventional feedforward controller if the learning parameters (i.e. ηs and αs set for the delta rule) are set to zero, thus providing a no-learning output of the system.

[2] The activation function is chosen to be a sigmoid of the form,

$$S_i = F_a^i \left(F_c^i, \beta_i \right) = \frac{1}{1 + e^{\left(\beta_i - F_c^i \right)}} - \frac{1}{2}$$

with its output range being within the interval [–0.5,0.5], thus accommodating for any negative results in the output data. Although other activation functions are possible, this form of the sigmoid is adequate for this work.

[3] All bias values are set to zero, thus $\beta = 0$ in the above equation of the sigmoid.

[4] The combining function is intended to form a relation between all inputs to a specific node and their weights. No restrictions on the form of such a function have been made, although a simple additive function of the form $\Sigma W.I$ is usually assumed. For this application, the combining function is taken to have a non-linear nature, and is set as one part of the system model for each neurone in the network. Hence, for each node,

$$F_c^i = f(P),$$

where $P = S_j^{-1}(J_1, J_2, \ldots, J_j)$, $J_j = I_i W_{ij}$ and S_j^{-1} is the inverse-sigmoid function defined as $S_j^{-1} = \beta_i - \ln[(1/(F_a^i + 1/2)) - 1]$. The combining function will take the form of a number of multiplicative and additive units arranged in a certain way, and will be different for each node in the network. Hence, the main task of each node is to execute one part of the mathematical model, and transmit its output to the nodes of the next hidden layer requiring it.

It should be mentioned, however, that one independent work reported the possibility of using sigma-pi functions of the form $\Sigma \prod W.I$ which were seen as the most complex pattern of connections ever needed [19]. None the less, this work develops multiple sigma-pi like functions which are certainly far more complex.

4.5 ROBOT CONTROL MODULES

A general form of the robot control loop [24,25] is shown in Figure 4.4, where the required task is first divided by the task planner producing a number of consecutive tasks, followed by the motion planner which gives a time history of positions, velocities and accelerations sufficient and necessary to realise each task. Once the desired motion elements are available, they are used to produce the commands for the individual joint loops via the control module which may or may not include the dynamic model of the system (model reference adaptive controllers vs. simple PIDs). The motion is realised by applying the control commands to the robot system and a feedback module provides the actual motion elements to cater for any uncertainties and/or changes in the system parameters and/or environment set-up. Overall, intelligence may be needed at different parts of the control loop, e.g. in connection with the task planner, dynamic model or sensory feedback. None the less, for complicated multi-joint mechanical chains, the above loop is very difficult to accomplish in real time and provisions must be made for efficient designs. The blocks shown in Figure 4.4 are from the view point of an upper hierarchical structure and serve to illustrate the main modules; both the intelligence and feedback blocks can be deployed in other parts of the system as an implementation may require.

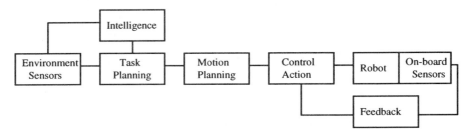

Figure 4.4 Overview of a robot integrated system

Two procedures are of particular importance. The *robot Jacobian* deals with the geometry and motion of the robot while considering the links' linear and angular velocities, while the robot dynamics takes into account the forces required to cause the motion and their effect on the system behaviour. Consequently, the *inverse Jacobian* formulation provides for joint rates as a feedback in closed-loop control while the *inverse dynamics* formulation computes the required torque (force) values given the desired parameters of motion (i.e. position, velocity and acceleration).

In this chapter, two robot modules are implemented, namely the inverse Jacobian and the inverse dynamics, to demonstrate the applicability and efficiency of the proposed approach. In addition, both modules are formulated using the full model of the Unimation PUMA 560 manipulator with six revolute joints.

4.5.1 Neural dynamic control (NDC)

One main issue in robot control is to solve for the dynamic non-linear equations of motion [24]. This is particularly important when the arm's speed and/or payload vary considerably while executing a pre-planned motion, hence leading to a change in the model. Therefore, the generalised torques/forces must be computed for all joints of the arm at each control cycle, which involves the solution of the inverse dynamics procedure. The Euler-Lagrange dynamic model of the robot is defined as follows:

$$\tau^i_{total} = \sum_{j=1}^{N} D_{ij}\ddot{\theta}_j + \sum_{j=1}^{N}\sum_{k=1}^{N} H_{ijk}\dot{\theta}_j\dot{\theta}_k + C_i, \quad i = 1,2,\ldots,N \tag{4.1}$$

where, $\tau^i \equiv$ torque value for joint $i, \theta, \dot{\theta}, \ddot{\theta} \equiv$ position, velocity and acceleration of each joint, $D_{ij} \equiv$ effective and coupling inertias, $H_{ijk} \equiv$ centripetal and coriolis forces, $C_i \equiv$ gravity loading, and $N \equiv$ number of joints of a given manipulator.

The computations of equation (4.1) are very difficult to attempt in real time. In addition, any simplifications in the model used will affect the accuracy of motion unacceptably [26]. Therefore, the principles of distributed processing offered by the artificial neural network concepts are investigated as a possible solution to the problem. In addition, the ability to change the network weights would provide for a learning control algorithm that may be adequate for real-time robotic applications.

4.5.1.1 Symbolic representation of the dynamic equations

The Euler-Lagrange approach to the formulation of the dynamic equations stated above is quite structured, with the torque value of each joint being expressed as the sum of three main terms, namely, the acceleration-related term τ_a, the velocity-related term τ_v, and the gravity term τ_g. Hence,

$$\tau_a^i = \sum_{j=1}^{N} D_{ij} \ddot{\theta}_j \tag{4.2}$$

$$\tau_v^i = \sum_{j=1}^{N} \sum_{k=1}^{N} H_{ijk} \dot{\theta}_j \dot{\theta}_k \tag{4.3}$$

$$\tau_g^i = C_i \tag{4.4}$$

yielding,

$$\tau_{total}^i = \tau_a^i + \tau_v^i + \tau_g^i \tag{4.5}$$

One method of representing these equations is in a symbolic form [27]. In such an approach, all unnecessary computations at lower levels of the formulation are omitted (e.g. multiplications by 1s and 0s, additions to 0s and common terms), thus producing a more compact code. In addition, all programming structures would be eliminated, reducing a large amount of unnecessary execution time. It should be noted, however, that, due to different configurations of serial-links arms, a symbolic representation should be produced for each robot arm individually. Several symbolic representations have been described in the literature for the Euler-Lagrange formulation of the PUMA manipulator [28,29].

Expanding the individual terms of equation (4.5) for a six-joint manipulator, the following representation could be found for each of the relevant terms.

For computing τ_a:

$$D = \begin{bmatrix} D_{11} & D_{12} & D_{13} & D_{14} & D_{15} & D_{16} \\ & D_{22} & D_{23} & D_{24} & D_{25} & D_{26} \\ & & D_{33} & D_{34} & D_{35} & D_{36} \\ & & & D_{44} & D_{45} & D_{46} \\ & \textit{Symmetric} & & & D_{55} & D_{56} \\ & & & & & D_{66} \end{bmatrix} \tag{4.6}$$

For computing τ_v:

$$H_1 = \begin{bmatrix} 0 & h_{112} & & h_{113} & h_{114} & h_{115} & h_{116} \\ & h_{122} & & h_{123} & h_{124} & h_{125} & h_{126} \\ & & & h_{133} & h_{134} & h_{135} & h_{136} \\ & & & & h_{144} & h_{145} & h_{146} \\ & \multicolumn{3}{c}{Symmetric} & & h_{155} & h_{156} \\ & & & & & & h_{166} \end{bmatrix} \tag{4.7}$$

$$H_2 = \begin{bmatrix} (h_{122}) & 0 & & h_{213} & h_{214} & h_{215} & h_{216} \\ & 0 & & h_{223} & h_{224} & h_{225} & h_{226} \\ & & & h_{233} & h_{234} & h_{235} & h_{236} \\ & & & & h_{244} & h_{245} & h_{246} \\ & \multicolumn{3}{c}{Symmetric} & & h_{255} & h_{256} \\ & & & & & & h_{266} \end{bmatrix} \tag{4.8}$$

$$H_3 = \begin{bmatrix} (-h_{113}) & (-h_{213}) & & 0 & h_{314} & h_{315} & h_{316} \\ & (-h_{223}) & & 0 & h_{324} & h_{325} & h_{326} \\ & & & 0 & h_{334} & h_{335} & h_{336} \\ & & & & h_{344} & h_{345} & h_{346} \\ & \multicolumn{3}{c}{Symmetric} & & h_{355} & h_{356} \\ & & & & & & h_{366} \end{bmatrix} \tag{4.9}$$

$$H_4 = \begin{bmatrix} (-h_{114}) & (-h_{214}) & & (-h_{314}) & 0 & h_{415} & h_{416} \\ & (-h_{224}) & & (-h_{324}) & 0 & h_{425} & h_{426} \\ & & & (-h_{334}) & 0 & h_{435} & h_{436} \\ & & & & 0 & h_{445} & h_{446} \\ & \multicolumn{3}{c}{Symmetric} & & h_{455} & h_{456} \\ & & & & & & h_{466} \end{bmatrix} \tag{4.10}$$

$$H_5 = \begin{bmatrix} (-h_{115}) & (-h_{215}) & & (-h_{315}) & (-h_{415}) & 0 & h_{516} \\ & (-h_{225}) & & (-h_{325}) & (-h_{425}) & 0 & h_{526} \\ & & & (-h_{335}) & (-h_{435}) & 0 & h_{536} \\ & & & & (-h_{445}) & 0 & h_{546} \\ & \multicolumn{3}{c}{Symmetric} & & 0 & h_{556} \\ & & & & & & h_{566} \end{bmatrix} \tag{4.11}$$

$$H_6 = \begin{bmatrix} (-h_{116}) & (-h_{216}) & (-h_{316}) & (-h_{416}) & (-h_{516}) & 0 \\ & (-h_{226}) & (-h_{326}) & (-h_{426}) & (-h_{526}) & 0 \\ & & (-h_{336}) & (-h_{436}) & (-h_{536}) & 0 \\ & & & (-h_{446}) & (-h_{546}) & 0 \\ & \textit{Symmetric} & & & (-h_{556}) & 0 \\ & & & & & 0 \end{bmatrix} \qquad (4.12)$$

For computing τ_g:

$$C = \begin{bmatrix} c_1 & c_2 & c_3 & c_4 & c_5 & c_6 \end{bmatrix}^T \qquad (4.13)$$

The symbolic representation for each term shown in equations (4.6)-(4.13) computed for the PUMA with a particular gripper and a certain load [29].

4.5.1.2 The distributed formulation

The form of equations (4.6)-(4.13) readily lends itself to a distributed recasting, introducing certain concepts of parallel processing. In addition, all terms shown in parenthesis are redundant and can be obtained from previously computed elements of the equations. However, using a general purpose multi-processing system would introduce tremendous difficulties for a MIMD structure. In practical implementations, the communications overhead, which is certainly extensive in this case, could severely disrupt the efficiency of the distributed algorithm. Although theoretical formulations do not usually take proper account of this problem [30], it has been proven to have its effect on actual multi-processor systems [27].

Nevertheless, the use of specialized VLSI structures promises to present a solution to such a problem. If each of these terms were to be placed on separate computational nodes, each having the ability of intercommunication, and, further, a suitable efficient distributed algorithm was to be provided, satisfactory results could be expected. The computational nodes required for each term of the dynamic equations are twenty nodes for τ_a, sixty-seven nodes for τ_v, and five nodes for τ_g.

4.5.2 Neural inverse Jacobian (NIJ)

The most computationally efficient formulation for the inverse Jacobian for a manipulator with a spherical wrist is given as [31]

$$\dot{\Theta} = J^{-1}\dot{\chi} \qquad (4.14)$$

with J^{-1}, the inverse Jacobian matrix, defined as,

$$J^{-1} = \begin{bmatrix} J_{11} & 0 \\ J_{21} & J_{22} \end{bmatrix}^{-1} \qquad (4.15)$$

where $\Theta = [\theta_1,\theta_2,\theta_3,\theta_4,\theta_5,\theta_6]^T$ and $\chi = [P_X,P_y,P_Z,D_X,D_y,D_Z]^T$ are the joint rates and the Cartesian rates, respectively. Alternatively, J^{-1} can be stated as

$$J^{-1} = \begin{bmatrix} J_{11}^{-1} & 0 \\ -J_{22}^{-1} J_{21} J_{11}^{-1} & J_{22}^{-1} \end{bmatrix}$$ (4.16)

4.6 CONSTRUCTING THE NEURAL STRUCTURES

In order to construct a neural structure for both the formulations of Sections 4.5.1 and 4.5.2, the values of I_d and O_d must be identified for a start. For the NDC, I is set to $\{\Theta, \dot{\Theta}, \ddot{\Theta}\}$ and O is set to τ, while for the NIJ, $I \equiv \chi$ and $O \equiv \Theta$. For both formulations, the network would also acquire the manipulator's link parameters as inputs.

4.6.1 The NDC network

Since the total torque evaluation process could be subdivided into three main parts as indicated by equation (4.5), the neural network is to be constructed accordingly. Initially, an input layer of eighteen nodes is needed, where six nodes are used for each set of the parameters of motion (i.e. $\Theta, \dot{\Theta}, \ddot{\Theta}$). In addition, an output layer of six nodes is considered, yielding the required motor inputs for the six-joint robot. Three sets of hidden layers are designated for the distributed algorithm, each set accommodating for one term of the dynamic equations :

- Acceleration-related term: two hidden layers are required with twenty and six nodes in each, respectively.
- Velocity-related term: three hidden layers are required with sixty-seven, thirty-six and six nodes in each, respectively.
- Gravity term: only one hidden layer is required with five nodes.

The role of each set of hidden layers is to produce one part of the torque values, which are then collected and added together by nodes of the output layer. The structure of the proposed network is shown in Figure 4.5. The function of each of the hidden layers shown in Figure 4.5 is as follows :

- [1] Computing τ_a
- Hidden layer no. 1: each node of this layer computes one element of the inertial matrix (equation (4.6)).
- Hidden layer no. 2: each node multiplies one row of the inertial matrix (computed by the previous layer) by the acceleration vector of all joints, thus producing τ_a for each joint of the PUMA given by equation (4.3).

[2] Computing τ_v

- Hidden layer no. 1: each node of this layer computes one element of the Coriolis and centripetal force expressed by equations (4.7)-(4.12).
- Hidden layer no. 2: each node multiplies one row of a matrix of the Coriolis and centripetal forces related to one joint (computed by the previous layer) by the velocity vector of all joints.

- Hidden layer no. 3: each node multiplies the output vector computed by several nodes of the previous layer times the velocity vector of all joints, thus producing τ_v for each joint of the PUMA given by equation (4.4).

[3] Computing τ_g

- Hidden layer no. 1: each node of this layer computes one element of the position-related gravity term expressed by equation (4.13).

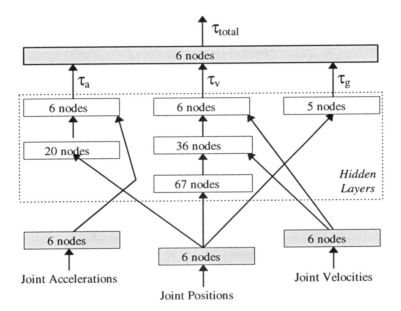

Figure 4.5 The NDC neural structure

The connections amongst nodes throughout the network is sparse, depending upon the needs of the mathematical formulation. The complete neural network is shown in Figure 4.6 for all parts of the computations.

4.6.2 The NIJ network
The distributed formulation proposes setting different terms (or operations) of equations (4.15) and (4.16) on separate computational elements representing the nodes in the multi-layered network. Each of these terms acts as the combining function for a particular node. Thus, the number of hidden layers, the number of nodes in each hidden layer and the connections between different layers will be imposed by the mathematical formulation.

Thus, a total of six hidden layers are needed, as shown in Figure 4.7. The function of each of these hidden layers is as follows:

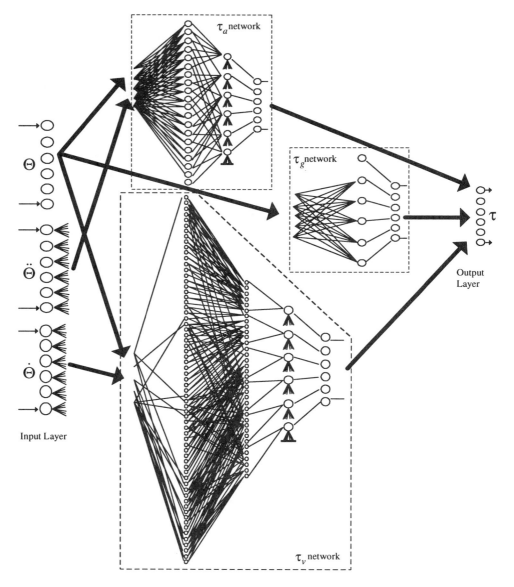

Figure 4.6 The NDC complete neural network

- Hidden layer no. 1: computes the elements required for the third column of J_{11} and the first column of J_{22}.
- Hidden layer no. 2: computes elements for the first and second columns of J_{11} and the first and third columns of J_{21}.

- Hidden layer no. 3: computes the complete J matrix and elements of J_{22}^{-1}.
- Hidden layer no. 4: computes J_{11}^{-1}.
- Hidden layer no. 5: computes $-J_{22}^{-1} J_{21} J_{11}^{-1}$.
- Hidden layer no. 6: computes the required joint rates by multiplying the resultant inverse Jacobian and the given Cartesian rates.

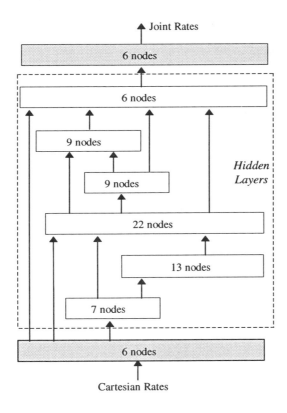

Figure 4.7 The NIJ neural structure

The above structure is dictated by the very formulation of the inverse Jacobian procedure, where nodes are computed whenever their pre-required elements are available, thus making their outputs available for the nodes of the next layers. Hence, while the computation of J_{22}^{-1} does not depend on the total evaluation of J, both are computed in one layer. The complete neural network is illustrated in Figure 4.8.

4.7 SIMULATION RESULTS

In order to test the developed neural structure, case studies are programmed and simulated for both the NDC and the NIJ structures. Simulation results are included for pre-defined motions of a PUMA 560 robot arm.

4.7.1 NDC simulations

As a case study, six-joint trajectories (one for each joint of the PUMA) have been considered, for which a total of 1500 control cycles are executed. Referring to Figure 4.3, the robot arm can be simulated using the direct dynamics algorithm [32], which produces the actual parameters of the motion executed by the arm.

To demonstrate the ability of the neural controller to follow a change of robot model on-line, the arm's payload was changed during the motion, thus producing significant tracking errors, as indicated by the first row of Table 4.1. A change in the payload from 4.22 kg to 4.00 kg was imposed after completion of 25% of the desired motion (i.e. 375 control cycles). However, due to the learning capabilities, the maximum tracking errors were reduced significantly, as shown by the second row of Table 4.1.

Figure 4.9 shows the simulated motion for the uncontrolled and controlled trajectories of joints 1, 2 and 3, respectively, indicating the position error reduction and its corresponding torque changes. The term uncontrolled trajectory is used to indicate the motion executed without the neural controller. Similar performances can be observed for the other three joints of the arm. The chosen learning and momentum factors are shown in Table 4.2, while the adjusted weights of layer no. 1 of the acceleration-related term are shown in Table 4.3 for illustration.

The total required execution time of the distributed network is 5.85 milliseconds (programmed in C and run on a Sun 4), which corresponds to the time taken to compute element D_{11}, in addition to the time required for the computation of one node of layer no. 2 of the acceleration-related term, layer no. 3 of the velocity-related term, and the output layer.

Table 4.1 Reducing the errors (NDC simulations)

PUMA Joint	Tracking Errors (degrees)					
	1	2	3	4	5	6
No Learning	4.823	2.566	3.626	55.93	64.38	45.16
With Learning	2.933	0.876	1.684	37.08	50.81	34.22

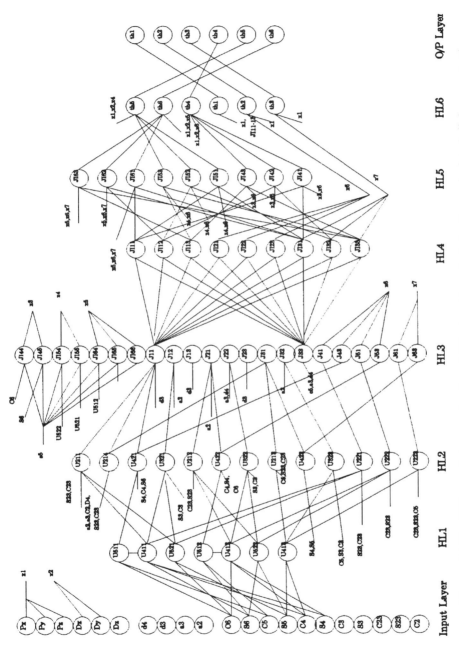

Figure 4.8 The NIJ complete neural network (The input layer includes the manipulator's link parameters)

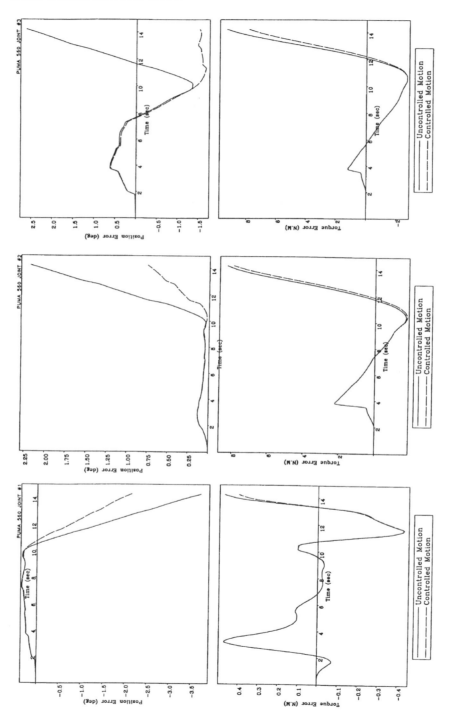

Figure 4.9 Motion tracking of joints 1, 2 and 3 (NDC simulations)

Table 4.2 Learning and momentum factors (NDC simulations)

	Layer No.	Learning (η)	Momentum (α)
Acceleration-related	1	0.0001	0.5
	2	0.0001	0.6
	1	0.0002	0.5
Velocity-related	2	0.0002	0.5
	3	0.0002	0.5
Gravity	1	0.0002	0.5

4.7.2 Neural inverse Jacobian simulation

In this simulation, a Cartesian trajectory has been specified for which a total of 900 control cycles are executed. To demonstrate the ability of the NIJ to adapt on-line for changes in the robot model, one of the arm's parameters (d_3=0.1254 m) was changed after completing 25% of the specified motion. As a result, tracking errors were detected. However, applying the learning algorithm led to considerable reduction in the deviation errors, as shown in Table 4.4. In addition, Figure 4.10 indicates the controlled and uncontrolled motions of the first three joints of the PUMA. All joint errors were reduced significantly except for joints 4 and 6, which is attributed to being caught in a local minima.

For each node in the network, the learning factor (η) is set to 0.005 while the momentum factor (α) ranges between 0.48 and 0.51. Computing this simulation of a similar system to that used for Section 4.7.1, the heaviest computational node was no. 2 of layer no. 5 which required 1.93 milliseconds to evaluate.

4.8 DISCUSSION

In a general control system, the tasks of the neural-based controller are as follows:
1. Identification of the system model: Although no specific methodologies are presented, it has been suggested that the interaction between the controller and some system would eventually lead to the determination of the appropriate model of the latter.
2. Estimation of the model changes: The interaction of the neural controller with the system leads to the observation of any differences between the two, thus producing a more accurate model.
3. Controlling the system: depending on the output of the estimation procedure.

These tasks would allow for the on-line adaptation of the model mapped on the neural network to any changes in the controlled system or in its surrounding environment.

The example of the animal brain was given as a realistic implementation of the above procedure, where the brain would accommodate by repetitive experiments for any physical changes in the animal body, along with any related changes in the environment.

Table 4.3 Adjusted weights of the acceleration-related term
(first hidden layer)

Node #	Weight #				
	W_1	W_2	W_3	W_4	W_5
1	1.000082	0.998636	0.999273	1.000407	0.999370
2	0.999349	1.001699	1.001647	0.998983	1.001070
3	1.000177	1.002572	1.000791	0.999768	1.000916
4	0.999990	1.000159	1.000023	1.000091	1.000001
5	0.999741	1.001401	1.000949	0.999497	1.000697
6	0.999983	0.999547	0.999830	1.000091	0.999820
7	0.999674	0.999813	1.000101	0.999847	NA †
8	0.996843	0.998869	1.000444	0.998810	NA
9	1.001862	1.000829	0.999439	1.000849	NA
10	1.004906	1.001438	0.999455	1.001765	NA
11	0.999870	0.999962	1.000022	0.999946	NA
12	0.998655	1.000726	0.998935	NA	NA
13	1.002068	0.998751	1.001625	NA	NA
14	1.000420	0.999986	1.000845	NA	NA
15	0.999770	1.000115	0.999826	NA	NA
16	0.999948	1.000249	NA	NA	NA
17	0.999924	1.000079	NA	NA	NA
18	0.999992	NA	NA	NA	NA
19	1.000640	NA	NA	NA	NA
20	NA	NA	NA	NA	NA
† NA = Not Applicable					

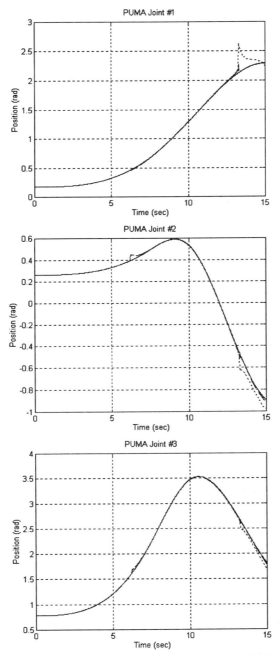

Figure 4.10 Motion tracking of the first three joints (NIJ simulations)
(___ desired motion; .... actual uncontrolled motion; ------ controlled motion)

Table 4.4 Reducing the errors (NIJ simulations)

PUMA Joint	Tracking Errors (degrees)					
	1	2	3	4	5	6
No Learning	0.511	0.159	0.164	0.264	0.457	0.251
With Learning	0.079	0.052	0.052	0.251	0.065	0.247

However, this example does not explain how the initial parametric representation of the model can be constructed.

In the case of robot manipulators, the non-linear system has proved to be very complicated for expression by arbitrarily constructed layers of an experimental number of neurones, as reported in the related literature. Therefore, it is concluded that including an initial form of the robot model within the neural controller would give a starting point for its learning behaviour. A controller is expected to perform more efficiently when it contains some knowledge of the controlled system. Hence, the dynamic equations of the robot arm have been included in the neural network, as described in previous sections.

Once an initial form of the system model is available for the neural controller, the estimation of changes can be based on the actual behaviour of the system, and the model parameters can be further adapted to force the system to behave in the desired way.

The main aim of the proposed neural model has been twofold. First, cutting down the computational complexities by introducing a distributed algorithm for the involved equations. Second, providing a learning scheme through which adaptation to any changes in the environment could be accommodated.

As the proposed neural structure will be set according to the mathematical model of a system, the number of nodes, and indeed layers, can be increased by further division of the combining function. This is illustrated in Figure 4.11 for the hypothetical case of a massive distribution of element D_{11} of equation (4.6). In Figure 4.11, the initial node containing the D_{11} element as its combining function can be replaced by a number of nodes with combining functions comprising parts of the D_{11} element, hence leading eventually to massive parallelism. Further work is in progress to develop a methodology for realising such a procedure.

4.9 CONCLUSIONS

A new approach to neural-based control has been presented, where conventional control theory is augmented with aspects of the theory of cognition. This approach gives the neural network some initial knowledge of the system model, where the network is expected to behave more efficiently once it knows something about the task it is required to perform. Such initial knowledge can be readily available in engineering applications where data of the controlled system is provided for, as is the case in robotics.

The main aim of this approach is to make use of the parallelism for real-time implementation, in addition to the learning abilities to perform adaptive control. The

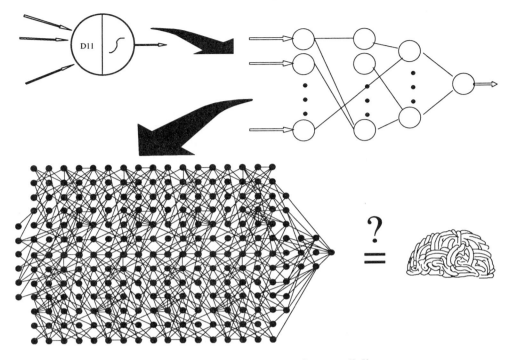

Figure 4.11 Introducing massive parallelism

implementations reported illustrate the practicality of the procedure for two modules used in robot control, considering any robot structure with any number of joints. In addition, no reservations are made for the type of executed motion.

REFERENCES

[1] Slotine, J.J.E. and Gunter, N. (1991) Performance in adaptive manipulator control, *Int. J. Robotics Research*, Vol.10, No.2.

[2] Zbikowski, R. and Gawthrop, P.J. (1992) A survey of neural networks for control. In *Neural networks for control and systems*, eds.: K. Warwick, G.W. Irwin, K.J. Hunt, IEE, London.

[3] Kawato, M.Y., Uno, Y., Osobe, M. and Suzuki, R. (1988) Hierarchical neural network model for voluntary movement with application to robotics, *IEEE Control Systems Magazine*, pp 8-16.

[4] Ruano, A.E.B., Jones, D.I. and Fleming, P.J. (1991) A neural network controller, In *Proc. IFAC Workshop on Algorithms and Architectures for Real-Time Control*, Bangor, UK.

[5]   Ziauddin, S.M. and Zalzala, A.M.S. (1994) Practical parallel processing for real-time adaptive control of mechanical arms. In *Proc. IEEE Int. Conf. on Control Applications*, pp.315-20, Strathclyde, UK.

[6]   Billings, S. and Chen, S. (1992) Neural networks and system identification. In *Neural networks for control and systems*, eds.: K. Warwick, G.W. Irwin, K.J. Hunt, IEE, London.

[7]   Bank, S.P. and Harrison, R.F. (1989) Can perceptrons find Lyapunov functions? - an algorithmic approach to systems stability, *Research Report #365,* Dept. of Automatic Control Systems Engineering, University of Sheffield, UK.

[8]   Pao Y.-H. (1989) *Adaptive pattern recognition and neural networks*, Addison Wesley.

[9]   Berhen, J., Gulati, S. and Zak, M. (1989) Neural learning of constrained non-linear transformations, *IEEE Computer*, Vol. 22, No. 6, pp. 67-76.

[10] Roele, M. and Warwick, K. (1991) Combining adaptive and neural control. In *Proc. IFAC Workshop on Algorithms and Architectures for Real-Time Control*, Bangor, UK.

[11] Arbib, M.A. (1989) *The metaphorical brain 2: neural networks and beyond*, Wiley.

[12] Pao, Y.H. and Sobajic, D.J. (1987) Artificial neural-net based intelligent robotic control. In *Proc. SPIE Robotics/IECON 87 Meeting*, pp. 542-9.

[13] Sobajic, D.J., Lu, J.J. and Pao, Y.H. (1988) Intelligent control of the Intelledex 605T robot manipulator. In *Proc IEEE Int. Conf. Neural Networks, Vol.2*, pp. 433-40.

[14] Guez, A. and Ahmad, Z. (1988) Solution to the inverse kinematics problem in robotics by neural networks. In *Proc. IEEE Int. Conf. Neural Networks, Vol.2*, pp. 617-23.

[15] Jorgensen, C.C. (1987) Neural network representation of sensor graphs in autonomous robot path planning. In *Proc. IEEE Int. Conf. Neural Networks, Vol.4*, pp. 507-15.

[16] Guez, A., Eilbert, J. and Kam, M. (1987) Neuromorphic architecture for adaptive robot control: A preliminary analysis. In *Proc. IEEE Int. Conf/ Neural Networks, Vol.4*, pp. 567-72.

[17] Ritter, H.J., Martinez, T.M. and Schulten, K.J. (1989) Topology-conserving maps for learning visui-motor-co-ordination, *Neural Networks, Vol.2*, pp. 159-68.

[18] Lippmann, R.P. (1987) An introduction to computing with neural nets, *IEEE ASSP Magazine*, pp. 4-22.

[19] McClelland, J.L. and Rumelhart, D.E. (1988) *Explorations in parallel distributed processing,* MIT Press.

[20] Crick, F.H. and Asanumna, C. (1986) Certain aspects of the anatomy and physiology of the cerebral cortex. In *Parallel, Distributed Processing: Explorations in the Microstructure of Cognition*, ed. J.L. McClelland, Vol.2, pp. 333-40, MIT Press.

[21] Blum, A.L. and Rivest, R.L. (1992) Training a 3-node neural network is NP-complete, *Neural Networks*, Vol.5, pp. 117-27.

[22] Rumelhart, D.E., Hinton, G.E. and Williams, R.J. (1986) Learning representations by backpropagating errors, *Nature,* Vol.323, pp. 533-6.

[23] Minsky, M.L. and Papert, S.A. (1988) Epilog: the new connectionism. In *Perceptrons*, pp. 247-80, MIT Press.

[24] Fu, K. S., Gonzalez, R.C. and Lee, C.S.G. (1987) *Robotics: Control sensing, vision and intelligence*. McGraw-Hill, New York.

[25] Paul, R.P. (1981) *Robot Manipulators: Mathematics, programming and control*, MIT Press.

*[26]*Hollerbach, J.M. (1984) Dynamic scaling of manipulator trajectories, *Trans. ASME, J. Dyn. Sys., Meas. and Control,* Vol.106, pp. 102-6.

[27] Zalzala, A.M.S. and Morris, A.S. (1989) A distributed pipelined architecture of the recursive Lagrangian equations of motion for robot manipulators with VLSI implementation, *Research Report #353,* Dept. of Automatic Control and Systems Eng., University of Sheffield, UK.

[28] Neuman, C.P. and Murray, J.J. (1987) The complete dynamic model and customized algorithms of the PUMA robot, *IEEE Trans. Syst. Man., Cyber.,* Vol.17, No.4, pp. 635-44.

[29] Tarn T.J., Bejczy, A.K., Yun, X. and Ding, X. (1986) Dynamic equations for six-link PUMA 560 robot arm, *Robotics Laboratory Report SSM-RL-85-05,* Dept. of Systems Science and Mathematics, Washington University.

[30] Luh, J.Y.S. and Lin, C.S. (1982), Scheduling of parallel computation for a computer controlled mechanical manipulator, *IEEE Trans. Syst., Man., Cyber.,* Vol.17, No.2, pp.214-34.

[31] Paul, R.P. and Zhang, H. (1986) Computationally efficient kinematics for manipulators with spherical wrists based on the homogeneous transformation representation, *Int. J. Robotics Research,* Vol. 5, No. 2, pp. 32-44.

[32] Walker, M.W. and Orin, D.E. (1982) Efficient dynamic computer simulation of robotic mechanisms, *Trans. ASME, J. Dyn. Syst. Meas. & Cntrl,* Vol.104, pp. 205-11.

5

Intelligent co-ordination of multiple systems with neural networks

Xianzhong Cui and Kang G. Shin
Real-Time Computing Laboratory, Department of Electrical Engineering and Computer Science, University of Michigan, Ann Arbor, MI 48109-2122, USA.

ABSTRACT

Many control applications require the co-operation of two or more independently designed, separately located, but mutually affecting, subsystems. In addition to the good behaviour of each subsystem, effective co-ordination of these subsystems is very important in order to achieve the desired overall system performance. However, such co-ordination is very difficult to accomplish due mainly to the lack of precise system models and/or dynamic parameters, as well as the lack of efficient tools for system analysis, design and real-time computation of optimal solutions.

In this chapter, we propose a new multiple-system co-ordinator which combines the techniques of intelligent control and neural networks, and forms the high-level co-ordinator in a hierarchical structure. The basic idea is to estimate the effects of the control commands to subsystems using a predictor, and modify these commands using a knowledge-based co-ordinator so as to achieve the desired performance. The predictor is designed for multiple-input, multiple-output systems using neural networks. The knowledge-based co-ordinator is responsible both for a goal-oriented search in its knowledge base and also for the overall system stability. Because the internal structure and parameters of the low level subsystems are not affected by using the proposed method, some commercially designed servo controllers for single systems can be co-ordinated and made to perform more sophisticated tasks for multiple systems than they were originally intended to do.

5.1 INTRODUCTION

Although some basic principles in co-ordinating multiple systems were developed in the early 1980s [1], most related publications addressed only conceptual interpretation, and very few of them dealt with actual applications. The main difficulty in co-ordinating multiple systems comes from the lack of precise system models and parameters, as well as the lack of efficient tools for system analysis, design and real-time computation of optimal solutions. New methods for analysis and design are thus required for the closed-loop co-ordination of multiple systems.

Since intelligent control does not depend only on mathematical analyses and manipulations, it is an attractive candidate for dealing with complex system control problems. An intelligent controller achieves the desired performance by searching for a goal in its knowledge base. There are three basic structures for intelligent control: performance-adaptive, parameter-adaptive, and hierarchical. The performance-adaptive structure is motivated by human expert control and/or human cognition ability, and attempts to control a system directly with an intelligent controller. Several examples of this structure are given in [4, 5, 6] and [8]. On the other hand, in a parameter-adaptive structure, the intelligent controller works as an on-line tuner of a conventional (usually PID) controller [6, 7, 8]. In a hierarchical structure, the intelligent controller [9] is a high-level controller which attempts to modify only the reference input to the low level subsystems. The low-level subsystem could be a servo control system, and its internal structure and parameters are not affected by adding this high-level controller. One of the main tasks associated with an intelligent controller is to design a knowledge base. An inference engine will then conduct a goal-oriented search in the knowledge base according to the characteristics of system performance. The error and/or error increment of system output, and the quality of a step response are commonly used to evaluate system performance. Additional characteristics have also been suggested. For example, the estimated, dominant pole location of a closed-loop system to express system performance was suggested in [5], though no knowledge base was built on it. In [4] the output error and its derivative were arranged into a phase plane divided into forty-eight areas, on the basis of which rules were designed. The goal was to control the system to reach the origin of this plane. In [9], the multiple-step prediction of system output was used to characterise system performance, and the knowledge for controlling the system was then simply represented by a decision tree.

However, all the results reported in the literature were intended for single systems. Most of the system characteristics mentioned above may not be suitable for co-ordinating multiple systems, because system performance may not be easily defined and related to the measured data and control inputs. In fact, for a complex system, even human knowledge on how to co-ordinate it to achieve the desired performance is limited and incomplete. Thus, it is difficult to design a complete knowledge base for such a system. Addition of a co-ordinator (not necessarily an intelligent one) leads to the problem of co-ordinating multiple systems to form a hierarchical structure. Such an addition should not interfere with the internal structure and parameters of low-level subsystems, making the structure of performance- or parameter-adaptive intelligent controllers unsuitable for multiple-system co-ordination. The internal structure and/or parameters of low-level subsystems are usually not known to the co-ordinator. Moreover, stability analysis becomes very important, due mainly to the uncertain low-level structure and/or parameters, and incomplete knowledge of the co-ordination and system characteristics. We should therefore ask the following questions when designing an intelligent co-ordinator.

1.   What are the strategy and the structure for co-ordinating multiple systems?
2.   What are the characteristics of multiple-system performance?
3.   What knowledge is necessary for co-ordination?
4.   How should the knowledge be represented?
5.   How can the qualitative knowledge be extracted from sensor data?
6.   How can the result of qualitative reasoning be changed into the quantitative control signals of actuators?
7.   How can system stability be analysed and guaranteed?

We propose a knowledge-based co-ordinator (KBC) for multiple systems by combining the techniques of intelligent control and neural networks (NNs). The KBC is a high-level co-ordinator within a hierarchical structure. The detailed structure and/or parameters of low-level subsystems are not required by the KBC, thus allowing individual subsystems to be designed independently. This implies that some commercially designed controllers can be co-ordinated to perform more sophisticated tasks than originally intended. In Section 5.2, the problem of multiple-system co-ordination is stated, and some basic principles of multiple-system co-ordination are reviewed. The proposed scheme and the assumptions used are described in Section 5.3. Section 5.4 addresses the design of a KBC, including the knowledge representation, solution, existence and system stability. Section 5.5 deals with the design of an NN-based predictor with multiple-input multiple-output (MIMO). The basic structure of the NN-based predictor, the updating problem and the training algorithm will be discussed there. As an example, the co-ordination of two 2-link robots holding a single object, and the co-ordination of two 2-link robots for collision avoidance are presented in Sections 5.6 and 5.7 respectively. The chapter concludes with Section 5.8.

5.2 PROBLEM AND PRINCIPLES OF MULTIPLE-SYSTEM CO-ORDINATION

Figure 5.1 describes two interacting systems, and this description can easily be generalised to the case of more than two systems. The system dynamics are described by

$$S_1(U_1, Y_1, W_2) = 0 \quad \text{and} \quad S_2(U_2, Y_2, W_1) = 0$$

where $U_i \in R^{n_i}, W_i \in R^{m_i}$, and $Y_i \in R^{p_i}$, for $i = 1, 2.$

Let $p = p_1 + p_2$, $n = n_1 + n_2$ and $m = m_1 + m_2$. The constraints are expressed by:

$$S_0 = \{(U,Y,W) \; : \; S_1 = 0, S_2 = 0\}$$

where $U = \left[U_1^T, U_2^T\right]^T \in R^n$ is the augmented control vector, $Y = \left[Y_1^T, Y_2^T\right]^T \in R^p$ is the augmented system output vector, and $W = \left[W_1^T, W_2^T\right]^T \in R^m$ is the vector representing interactions between the two systems.

Usually, the cost function of a multiple system is the sum of the cost functions of all component systems:

$$J(U,Y,W) \equiv J_1(U_1, Y_1, W_2) + J_2(U_2, Y_2, W_1) \tag{5.1}$$

The problem of co-ordinating multiple systems can be stated as an optimisation problem: *minimise* the cost function, J, subject to the constraint, S_0. Though there are no general approaches to solving this problem for a complex multiple system, some conceptual methods and basic principles have been suggested in [1]. One of these methods is called model co-ordination. Under this method, the problem is divided into two-level optimisation problems. First, suppose the interaction, W, is fixed at Z. Then compute

$$H(Z) = \min \left[J(U,Y,Z) \right] \qquad \text{for} \;\; (U,Y,Z) \in S_0$$

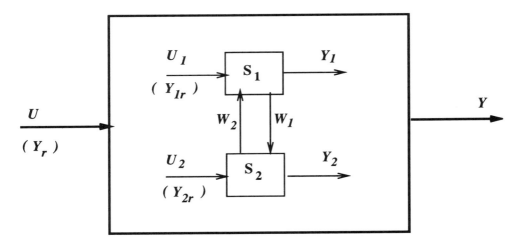

Figure 5.1 Interaction of two systems

$H(Z)$ is then minimised over all allowable values of Z. This two-level optimisation problem is solved iteratively until the desired performance is achieved. Another method is called goal co-ordination, in which the system is represented as in Figure 5.2. Suppose W_i is not necessarily equal to X_i. The overall optimality is achieved by sequentially optimising two subsystems, while treating W_i as an ordinary input variable of each corresponding subsystem. This requires X_i and W_i to be equal, which is called the interaction balance principle. Similar to the process of model co-ordination, optimality is achieved iteratively. Another basic principle of co-ordination, called the interaction prediction principle, is stated as follows. Let $\hat{W} = \left[\hat{W}_1^T, \hat{W}_2^T\right]^T$ be the predicted interaction and $W = \left[W_1^T, W_2^T\right]^T$ be the actual interaction under the control U. Then the overall optimum will be achieved if the prediction gives the true value, that is, $\hat{W} = W$.

Obviously, solving these optimisation problems largely depends on knowledge of the structure and/or the dynamic parameters of low-level subsystems and mathematical synthesis. Moreover, in a hierarchical system it is desirable that adding a high-level co-ordinator should not affect the internal structure and/or parameters of the low-level subsystems. The high-level co-ordinator should only give appropriate co-ordination commands to the low level subsystems, so that each level can be designed independently of other levels. That is, the higher the level, the more intelligence it has, and the less precise its knowledge about the low levels becomes. These requirements motivated us to design the KBC.

To design a co-ordinator, we first need to define a system performance index. This should be chosen to express the desired system performance and should also be amenable to some optimisation methods. For example, the performance index in Eq.(5.1) is suitable for the concepts of model co-ordination and goal co-ordination.

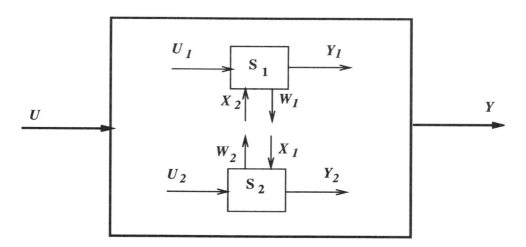

Figure 5.2 Goal co-ordination of two systems

To design a KBC, one needs an index to express system performance explicitly: such an index will henceforth be called the principal output. The overall system performance index may not necessarily be the simple summation of the performance indices of all component systems. Because only system constraints are important for co-ordination, one may not even be able to define subsystem performance indices. Moreover, we want to relate the principal output directly to the co-ordination commands. The co-ordination commands are defined as the reference to subsystems. The following sections will show that both the explicit expression for system performance and the direct relationship between the principal output and the co-ordination commands will simplify the design of the knowledge base and the goal-oriented search.

5.3 DESCRIPTION OF THE PRINCIPAL OUTPUT PREDICTION SCHEME

In a hierarchical structure, each level can be viewed as a mapping from its reference input to the output. The servo controller of each subsystem is usually designed separately from, and independently of, the others. In order not to interfere in the internal structure and/or parameters of the low level subsystems, the only effective control variable is the reference input to the lower level.

The reference inputs are a set of pre-designed commands, which represent the overall behaviour of the multiple system. For example, when multiple robots work in a common workspace, the reference input is the desired trajectory of each robot generated without considering the presence of other robots. The purpose of a high-level co-ordinator is to modify the desired trajectories to avoid collision among the robots. From a high-level co-ordinator's point of view, the following conditions are assumed:

C1   Each subsystem is a stable, closed-loop, controlled system.
C2   Each subsystem has a linear response to its reference input.
C3   Each subsystem will remain stable even during its interaction with other subsystems.
C4   System performance can be described explicitly by the principal output.

In Figure 5.1, let Y be the principal output vector of the multiple system. Also, let $Y_r = \left[Y_{1r}^T, Y_{2r}^T \right]^T$ be the vector of reference input to the low level subsystem. Note now that the components of Y may not simply be the outputs of the subsystems, but could be a function of these outputs:

$$Y = F_0 \left(Y_1, Y_2 \right), \qquad \text{where} \quad F_0 : R^{p1} \times R^{p2} \to R^p$$

Because each subsystem is a closed-loop system, Y_i can be represented as

$$Y_i = f_i(Y_{ir}, W_j), \qquad \text{where} \quad i, j = 1,2, \quad j \neq i, \quad \text{and} \quad f_i : R^{n_i} \times R^{m_j} \to R^{p_i}$$

Then Y can be represented as

$$Y = F(Y_{1r}, Y_{2r}, W_1, W_2) \tag{5.2}$$

where $F : R^{n_1} \times R^{n_2} \times R^{m_1} \times R^{m_2} \rightarrow R^p$. The principal-output vector Y in Eq.(5.2) establishes an explicit relationship between the overall system performance and the reference inputs.

Let $\hat{Y}(k+d/k)$ and $Y_d(k+d)$ be the d-step ahead prediction and the desired value of the principal output $Y(k)$ at time $k+d$, respectively. Then the performance index of the overall system can be defined as

$$J(k) = \left[Y_d(k+d) - \hat{Y}(k+d/k) \right]^T \left[Y_d(k+d) - \hat{Y}(k+d/k) \right]$$

The purpose of using a co-ordinator is to choose a suitable reference input vector $Y_r(k)$ in order to minimise $J(k)$ at time k subject to a set of constraints.

Suppose the prediction of the principal output corresponding to each choice of $Y_r(k)$ is available, and the constraints can be expressed by a set of production rules. Then, in each sampling interval, the desired performance can be obtained by iteratively trying different reference inputs and adjusting them according to the principal output prediction. For example, we propose the following algorithm to co-ordinate two subsystems, where the superscript i denotes the iteration count.

1. Compute the principal output prediction $\hat{Y}^0(k+d/k)$ for given reference inputs $Y_{1r}^0(k)$ and $Y_{2r}^0(k)$.

2. Using $\hat{Y}^i(k+d/k)$, modify the reference inputs of subsystem 1, $Y_{1r}^i(k)$, $i = 0, 1, 2, \ldots$.

3. Compute $\hat{Y}^{i+1}(k+d/k)$ for given reference inputs $Y_{1r}^i(k)$ and $Y_{2r}^0(k)$.

4. Set $i \leftarrow i+1$ and repeat steps (2) and (3) until $\hat{Y}^{i+1}(k+d/k)$ cannot be improved any further with $Y_{1r}^i(k)$ due to the constraints.

5. Set $i \leftarrow 0$.

6. Using $\hat{Y}^i(k+d/k)$, modify the reference inputs of subsystem 2, $Y_{2r}^i(k)$, $i = 0, 1, 2, \ldots$.

7. Compute $\hat{Y}^{i+1}(k+d/k)$ for given reference inputs $Y_{1r}^0(k)$ and $Y_{2r}^i(k)$.

8. Set $i \leftarrow i+1$ and repeat steps (6) and (7) until $\hat{Y}^{i+1}(k+d/k)$ cannot be improved any further with $Y_{2r}^i(k)$ due to the constraints.

9. Set $i \leftarrow 0$ and repeat steps (2) to (8) until $\hat{Y}^i(k+d/k)$ reaches its desired value.

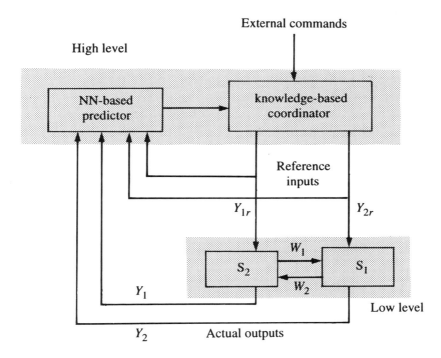

Figure 5.3 Conceptual structure of the knowledge-based co-ordination system

The conceptual structure of this scheme is given in Figure 5.3. Obviously, this scheme needs a multiple-step predictor to compute $\hat{Y}^i(k+d\,/\,k)$ and a KBC for the modification process of the reference inputs. By using this principal output predictor to characterise system performance, the knowledge for co-ordinating multiple systems becomes clear, thereby simplifying the design of a knowledge base.

We now need to address the following two problems: (1) given the principal output prediction, how can we design this KBC; and (2) how can we design such a principal output predictor? These two problems will be solved in the following sections.

5.4 DESIGN OF THE KNOWLEDGE-BASED CO-ORDINATOR

A multiple system with the KBC forms a hierarchical structure, and the low-level subsystems are viewed as a mapping from their reference input to the principal output. The goal is to modify the reference input so that the principal output reaches its desired value.

For a multiple system, we must define the principal output. Note that the knowledge-based co-ordination is not strictly a mathematical optimisation problem. The principal output must (1) have an explicit relation to the reference inputs, and (2) be measurable or computable from measured data. Because a multiple system is designed to perform a common task(s) among the component systems, such a principal output is usually defined to express the situation of the common task(s), even though it may not explicitly reflect some of the generally used optimisation criteria, such as energy or time.

As an example, consider the co-ordinated control of two robots. The two robots' operations may be tightly coupled or loosely coupled. They are tightly coupled, for example, when they hold a single object rigidly and are co-ordinated to move the object. On the other hand, they are loosely coupled when they work in a common workspace and are co-ordinated to avoid collision. Suppose each robot is equipped with a servo controller which was originally designed for a single robot. The two robots are co-ordinated by modifying each robot's reference input. For the tightly coupled case, the principal output can be defined as the object's position error or the internal/external force exerted on the object. For the loosely coupled case, on the other hand, the positions and/or velocities of the robot end-effectors can be used to represent the status of collision avoidance, and, thus, they are qualified to be the principal output. For both cases, an explicit relationship between system performance and reference input is established by defining the appropriate principal output.

As stated in the previous sections, we want to use the principal output predictor to see where each reference input of the subsystem will lead. If the principal output prediction is given, simplified knowledge on how to co-ordinate a multiple system can be stated in the following two steps:

1   Modify the reference input and feed the modified input to the predictor.
2   **IF** the principal output prediction yields good performance
      THEN feed the reference input to the low-level subsystems
      ELSE re-modify it.

Since only one reference input is modified at each time, the remaining problems are then: in which direction is the reference input modified (increase or decrease), how much should it be modified, and what are its limits? For a single system, we have developed such a knowledge-based controller in [9]. For a multiple system, the modification process of each reference input is similar to that of a single system, so only the related results of [9] are summarised below.

5.4.1 Knowledge representation

Using a predictor, the performance of a multiple system is characterised by the predicted tracking error in its principal output that results from the application of the current reference input. Thus, the space of predicted tracking errors forms the input space of the KBVC's knowledge base. The goal of the KBC is then to implement the modification process discussed thus far. It is not difficult to express this process in terms of a set of production rules.

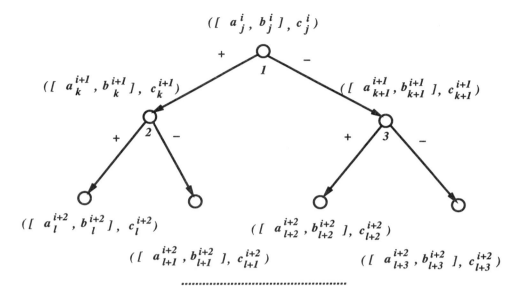

Figure 5.4 Decision tree

The possible actions that the KBC can take include increasing the reference inputs, decreasing the reference inputs, or keeping them unchanged. For each element of the reference input, the basic modification process can be represented by a decision tree, as shown in Figure 5.4. The *ij*th node in the tree is represented by $\left([a_j^i, b_j^i], c_j^i\right)$, where c_j^i is the quantity added to the reference input:[1]

$$y_r^{i+1}(k) = y_r^0(k) + c_j^i$$

$y_r^0(k)$ is an element of the original reference input vector to one of the subsystems at time k, y_r^{i+1} is its modified value after the *i*th iteration and $\left[a_j^i, b_j^i\right]$ is the interval to be searched, $a_j^i < c_j^i < b_j^i$ for all i, j.

---

[1]   Because only the reference input to one subsystem is modified at a time, in order to simplify the notation, subsystems 1 and 2 will not be distinguished within this section, that is, $y_r(k)$ will represent one element of either $Y_{1r}(k)$ or $Y_{2r}(k)$.

By giving the reference input $y_r^i(k)$ at any node $\left(\left[a_j^i, b_j^i\right], c_j^i\right)$, the interval $\left[a_j^i, b_j^i\right]$ is split into two subintervals $\left[a_k^{i+1}, b_k^{i+1}\right] \equiv \left[a_j^i, c_j^i\right]$ and $\left[a_{k+1}^{i+1}, b_{k+1}^{i+1}\right] \equiv \left[c_j^i, b_j^i\right]$, which form two successive nodes.

At the ith iteration and at the ijth node, let the predicted tracking error resulting from $y_r^i(k)$ be denoted as

$$e_j^i(k) = \hat{y}^i(k+d/k) - y_d(k+d),$$

where $y_d(k+d)$ is an element of $Y_d(k+d)$ and $\hat{y}^i(k+d/k)$ is the corresponding element of $\hat{Y}^i(k+d/k)$. Then, c_j^i is computed as:

$$c_j^i = \begin{cases} b_j^i - (b_j^i - a_j^i)K, & \text{if} \quad e_j^i(k) < 0 \\ a_j^i + (b_j^i - a_j^i)K, & \text{if} \quad e_j^i(k) > 0 \end{cases} \qquad (5.3)$$

and $0 < K < 1$ is a weighting coefficient which determines the step size of the iterative operation. a_0^0 and b_0^0 are the pre-designed lower and upper bounds for the amount of reference input modification, and usually $c_0^0 = 0$, that is, at the beginning, the reference input is not modified.

The structure of this decision tree shows that the simplest reference process is similar to forward chaining, starting from the root node. However, after a period of operation, we may learn that a positive augment c_j^i is always needed. Then, the inference process may start at any node with $c_j^i > 0$ and go forward or backward according to the sign of predicted tracking error. Note that this backward search does not mean a reverse search, but, rather, it means finding a suitable node to start the forward search. As soon as the forward search begins, the process is not reversible.

5.4.2 Solution existence and stability analysis
The basic forms of production rules are

IF $e_j^i(k) < 0$ AND $\left|e_j^i(k)\right| > \varepsilon$, THEN increase c_j^i AND compute $y_r^{i+1}(k) = y_r^0(k) + c_j^i$

IF $e_j^i(k) > 0$ AND $\left|e_j^i(k)\right| > \varepsilon$, THEN decrease c_j^i AND compute $y_r^{i+1}(k) = y_r^0(k) + c_j^i$

IF $\left|e_j^i(k)\right| \le \varepsilon$, THEN set $y_r^{i+1}(k) = y_r^i(k)$ AND stop the iterative operation

$\varepsilon > 0$ is a pre-specified error tolerance. Because the amount of modification to the reference input is bounded, or $a_0^0 < c_j^i < b_0^0$ for all i, j, there may be a case where $\left|e_j^i(k)\right| < \varepsilon$ for all c_j^i. This existence problem can be monitored by adding, for example, the following rule into the knowledge base, where δ is a pre-specified constant:

IF $\left(\left| c_j^i - b_0^0 \right| < \delta \quad \text{OR} \quad \left| c_j^i - a_0^0 \right| < \delta \right) \quad \text{AND} \quad \left| e_j^i(k) \right| > \varepsilon$

THEN change a_0^0 or b_0^0 automatically and continue the search
OR ask the operator for an adjustment
   OR stop the iterative operation and choose c_j^i with the smallest $e_j^i(k)$
     as the best output

Suppose the weighting factor K in Eq.(5.3) is set too small or too large, then the search for a proper c_j^i may take a very long time. This would not be acceptable if the required computation could not be completed within one sampling interval. The case of the computation/search time exceeding one sampling interval is equivalent to having no solution. This case is monitored by

IF the search time $> T_{\max}$ AND $\left| e_j^i(k) \right| < \varepsilon$
THEN stop the iterative operation
AND choose c_j^i with the smallest $e_j^i(k)$ as the best output
AND modify the weighting coefficient K

where $T_{\max}$ is a pre-selected allowable search time.

Suppose the prediction gives the true principal output, and let us consider the KBC and the closed-loop subsystem. The KBC can then be viewed as a map $M_0 : E \rightarrow Y_R$, specified by all the production rules, where E is the space of predicted principal-output tracking error and Y_R the reference input space. The low-level, closed-loop subsystem is also a map, $L : Y_R \rightarrow E$, which is specified by the desired dynamic properties of the servo controllers. Because L represents a well-designed controller and there exists a reference input at time k, $Y_r^i(k) \in Y_R$ such that the tracking error reaches zero. Accordingly, it is reasonable to assume that L is a linear map. The properties of the map $M \equiv L M_0 : E \rightarrow E$ depend mainly on the properties of the map M_0. In fact, all the antecedents of production rules are based on the prediction of principal output. If the predictor gives the true principal output, then the properties of the invariant map $M : E \rightarrow E$ are determined solely by the knowledge base.

For system stability, all production rules in the knowledge base must form a contraction map. More formally, we give the following theorem without proof (see [9] for its proof).

Theorem: Suppose (1) the principal-output prediction of a multiple-system is computable and the predictor gives the true principal output, and (2)$L : Y_R \rightarrow E$ of the low-level closed-loop subsystems is a linear map. If the map $M_0 : E \rightarrow Y_R$ is given by the decision tree, then we conclude that the composite map $M \equiv L M_0 : E \rightarrow E$ is a contraction map.

At each node of the decision tree, the iterative learning process is performed and the rules always keep the search direction pointed to the node where the tracking error decreases. This implies that the iterative learning process decreases the tracking error. Because the inference process is not reversible, it is impossible to have an unstable system response.

5.5 DESIGN OF AN NN-BASED PREDICTOR

Though it is assumed that the principal output, Y, is measurable or computable from the measured data, it may be very difficult to derive a closed-form expression for Eq.(5.2). Therefore, it is almost impracticable to design such a principal output using mathematical synthesis alone, even if such a closed form exists.

The development of NNs suggested that an I/O mapping can be approximated by a multi-layer perceptron [10,11]. With the ability of learning from examples, an NN can be trained to retain the dynamical property of an I/O mapping. Typically, a set of I/O pairs is arranged as $(u_1 , y_1), (u_2 , y_2),...,$ where $y_i = f(u_i)$ is a mapping. Using these training data, the connection weights within the NN are re-organised so as to represent the mapping relation. One of the most popular NN structures is the multi-layer perceptron with the backpropagation (BP) algorithm [12, 13]. The computation of BP includes two steps: (1) compute the NN's output forward from its INPUT to OUTPUT layers, and (2) modify the connection weights backward from its OUTPUT to INPUT layers.

We want to design a MIMO predictor using NNs. Referring to Eq.(5.2), the d-step ahead prediction of Y can be represented by

$$\hat{Y}(k + d / k) = F_p \ (\overline{Y}_{1r} , \overline{Y}_{2r} , \overline{Y}) \tag{5.4}$$

where:

$$\overline{Y}_{1r} = (Y_{1r} \ (k + i_1) ,..., Y_{1r} \ (k), Y_{1r} \ (k - 1),..., Y_{1r} \ (k - i_2))$$
$$\overline{Y}_{2r} = (Y_{2r} \ (k + j_1) ,..., Y_{2r} \ (k), Y_{2r} \ (k - 1),..., Y_{2r} \ (k - j_2))$$
$$\overline{Y} = (Y \ (k) , Y \ (k - 1) ,..., Y \ (k - i))$$
$$F_p \ : \ R^{n_1} \ \times \ R^{n_2} \ \times \ R^p \ \rightarrow \ R^p$$
$$i, i_1 , i_2 , j_1 \ \text{and} \ j_2 \ \text{are constant integers}$$

The interaction effects among subsystems are implicitly included in the historical data of $\overline{Y}$. In Eq.(5.4), the principal output prediction is directly represented as a mapping of the reference inputs and the historical data $\overline{Y}$. A three-layer (with one hidden layer) perceptron is designed to learn the relationship of Eq.(5.4), and forms the backbone of the NN-based predictor. Figure 5.5 shows the structure of the predictor, where the reference inputs

$\overline{Y}_{1r}$ and $\overline{Y}_{2r,}$ and the historical data $\overline{Y}$ are fed to the nodes at the INPUT layer. When the NN becomes well-trained, the predictions $\hat{Y}(k+d/k)$ for $d = 1, 2,...$ are then produced from the OUTPUT nodes. Two problems must be solved before implementing this NN-based predictor: (1) how to track a time-varying mapping; and (2) how to efficiently represent and compute a MIMO mapping with the NN. These problems have been treated in [14], and thus only the key points are summarised below for completeness.

5.5.1 Tracking a time-varying system

Suppose the predicted system is a SISO system with output y(k) and its d-step ahead prediction is $\hat{y}(k+d/k)$ at time k. Let $X_{2k}(k)$ and $X_{2k}^{d}(k)$ be, respectively, the actual output of the NN-based predictor and its desired value at time k. Then the network-output error is computed by

$$E_k(k) = X_{2k}^{d}(k) - X_{2k}(k)$$

In the standard BP algorithm, we must use this network-output error to train the NN. However, when the NN is used as a predictor, the network's output is the prediction of the system output, and the network-output error is the prediction error

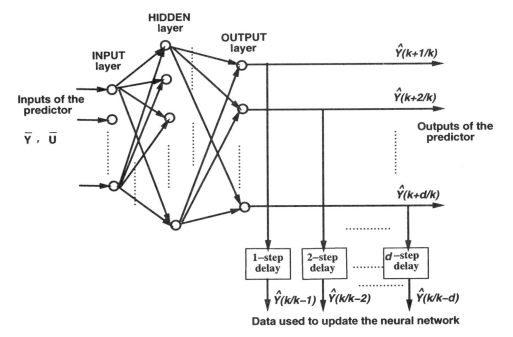

Figure 5.5 Structure of the NN-based predictor

$$E_k(k) = X_{2k}^d(k) - X_{2k}(k) = y(k+d) - \hat{y}(k+d\,/\,k) \tag{5.5}$$

$E_k(k)$ is unavailable since the system's future output $y(k+d)$ is not available at time k.

Hence, we must use the system's historical data to update the NN-based predictor on-line in order to maintain the closed-loop operation by keeping track of a time-varying system. To update the NN-based predictor, instead of using Eq. (5.5), we propose to use a posterior prediction error:

$$E_k(k-d) = X_{2k}^d(k-d) - X_{2k}(k-d) = y(k) - \hat{y}(k\,/\,k-d)$$

This arrangement is equivalent to cascading the NN with delay elements, as shown in Figure 5.5. We have derived a modified BP algorithm to handle these delay elements [14]. Therefore, the NN-based predictor is updated only by using the historical data of the predicted system and does not require knowledge of the dynamic parameters nor of the structure of the predicted system. Weights of the NN are dynamically adjusted to deal with the effects of nonlinear, time-varying properties, and/or long system-time delays. Because the NN-based predictor will always work in a closed loop, component failures in the NN will be learned and the NN will subsequently be re-configured. Furthermore, the parallel processing structure of the NN makes it suitable for high-dimensional systems.

5.5.2 Multi-dimensional back-propagation algorithm

Traditionally, to imitate an actual neuron, each node of an NN is usually designed to perform only scalar operations. However, for a MIMO system, if each node can only handle scalar operations, the size of the NN may become too large to manage. We therefore propose to equip each node of the NN with the ability to perform vector operations in order to easily specify some known coupling relations within the predicted system and to obtain an easier (thus more intuitive) form of the training algorithm.

All inputs and outputs of this NN are vectors. Referring to Figure 5.5, let $X_i \in R^n$, $X_{ij} \in R^m$, and $X_{2k} \in R^p$ be the output of the INPUT, HIDDEN and OUTPUT layer, respectively, for $1 \le i \le N, i \le j \le N_1$, and $1 \le k \le N_2$. The BP algorithm in a vector form has been derived in [14] and summarized below.

(i) Compute the output of the HIDDEN layer, X_{1j}

The output of the HIDDEN layer is computed by

$$X_{ij} \equiv \left[\, x_{1j1}, \ldots, x_{1jm} \,\right]^T = f_j\left(O_{ij}\right)$$

$$= \left[\frac{1}{1+\exp\left(-o_{1j1} - \theta_{1j1}\right)}, \ldots, \frac{1}{1+\exp\left(-o_{1jm} - \theta_{1jm}\right)} \right]^T \tag{5.6}$$

$$O_{ij} = \sum W_{ij}\, X_i, \quad j = 1, 2, \cdots, N_1 \tag{5.7}$$

where $W_{ij} \in R^{m \times n}$ is the weight matrix from node i of the INPUT layer to node j of the HIDDEN layer, $f_j : R^m \rightarrow R^m$ is defined as a sigmoid function of each component of a vector, and $\Theta_{ij} \equiv \left[\theta_{1j1}, \ldots, \theta_{1jm} \right]^T$ is the threshold vector at node j of the HIDDEN layer.

(ii) Compute the ouput of the OUTPUT layer, X_{2k}

The output of the OUTPUT layer is

$$X_{2k} \equiv \left[x_{2k1}, \ldots, x_{2kp} \right]^T = f_k (O_{2k})$$

$$= \left[\frac{1}{1 + \exp(-o_{2k1} - \theta_{2k1})}, \ldots, \frac{1}{1 + \exp(-o_{2kp} - \theta_{2kp})} \right]^T \qquad (5.8)$$

$$O_{2k} = \sum_{j=1}^{N_1} W_{1jk} \, X_{1j}, \quad k = 1, 2, \ldots N_2 \qquad (5.9)$$

where $W_{1jk} \in R^{p \times m}$ is the weight matrix from node j of the HIDDEN layer to node k of the OUTPUT layer, $f_k : R^p \rightarrow R^p$, is defined as a sigmoid function of each component of a vector, and $\Theta_{2k} \equiv \left[\theta_{2k1}, \ldots, \theta_{2kp} \right]^T$ is the threshold vector at node k of the OUTPUT layer.

Note that if $m=p=n$, $W_{ij} = \mathrm{diag}[w_{11}, \ldots, w_{nn}]_{ij}$ and $W_{1jk} = \mathrm{diag}[w_{111}, \ldots, w_{1nn}]_{jk}$, then the system is uncoupled.

(iii) Update the weights from the HIDDEN to OUTPUT layer, W_{1jk}

$$W_{1jk} (k+1) = W_{1jk} (k) + \Delta W_{1jk} \qquad (5.10)$$

where

$$\Delta W_{1jk} = \eta_1 \left[\delta_{1k} \, T_1 \right]^T$$

$$\delta_{1k} = (X_{2k}^d - X_{2k})^T \, \mathrm{diag} \left[x_{2k1} (1 - x_{2k1}), \ldots x_{2kp} (1 - x_{2kp}) \right]$$

T_1 is a $p \times m \times p$ tensor, with the lth matrix as

$$T_{1l} = \begin{bmatrix} 0 \\ \vdots \\ (X_{1j})^T \\ 0 \\ \vdots \end{bmatrix}^T \quad \leftarrow \text{ at the } l \text{ th row}$$

(iv) Update the weights from the INPUT to HIDDEN layer, W_{ij}

$$W_{ij}(k+1) = W_{ij}(k) + \Delta W_{ij} \tag{5.11}$$

where

$$\Delta W_{ij} = \eta \left[\delta_j \ T \right]^T$$

$$\delta_j = \left(\sum_{k=1}^{N_2} \delta_{1k} \ W_{1jk} \right) \text{diag} \left[x_{1j1} \left(1 - x_{1j1} \right), \ldots, x_{1jm} \left(1 - x_{1jm} \right) \right]$$

T_1 is an $m \times n \times m$ tensor, with the lth matrix as

$$T_l = \begin{bmatrix} 0 \\ \vdots \\ (X_i)^T \\ 0 \\ \vdots \end{bmatrix}^T \quad \leftarrow \int \quad at \ the \ l \ th \ row$$

(v) Update the thresholds at the OUTPUT and the HIDDEN layers

$$\Theta_{2k}(k+1) = \Theta_{2k}(k) + \eta_{1\theta} \left[\delta_{1k} \right]^T \tag{5.12}$$

$$\Theta_{1j}(k+1) = \Theta_{1j}(k) + \eta_\theta \left[\delta_j \right]^T \tag{5.13}$$

Extending the BP algorithm to a vector form shifts the complexity from the network level to the node level. Though the overall computation requirement is not reduced, it results in a set of succinct formulas and it is easier to specify the I/O nodes of the NN for an MIMO mapping and to express some known coupling relations. Moreover, if the NN is implemented in software and instructions for vector operations are provided, then the programming is more efficient with this vector form of BP algorithm.

With the ability for learning an input/output (I/O) mapping from experience, an NN can be used to track the variation of the mapping. However, an NN alone cannot form an intelligent co-ordination/control system. As a general method of representing systems with learning ability, NNs lack the ability for logical reasoning and decision making, for interpreting environmental changes, and for quick response to unexpected situations. Therefore, a KBC is needed. Despite its drawbacks, the NN-based predictor establishes an explicit relationship between the principal output and the reference inputs to subsystems. Hence, the knowledge base is simplified. One can also easily add to the knowledge base such rules as the constraints of subsystems, operation monitoring, system protection and switching of the co-ordination schemes. The KBC will emphasise system co-ordination but not data interpretation, while the ability to learn will rely mainly on the NN, that is, the NN will adapt itself to the model/parameter uncertainties, disturbances, component failures, and so on.

5.6   EXAMPLE 1: CO-ORDINATED CONTROL OF TWO 2-LINK ROBOTS

To demonstrate how to apply the proposed scheme to the solution of real-life problems, in this section we consider the problem of co-ordinating two 2-link robots[2] holding a rigid object, and leave an example of a loosely coupled system to the next section. The low-level subsystems include two robots, each with a separately designed servo controller. The basic configuration of this example is given in Figure 5.6 with the dynamic and kinematic parameters shown in Table 5.1.

Table 5.1   Kinematic and dynamic parameters of the simulated robot

	Length	Mass Center	Mass	Moment of Inertia
Link 1	1 m	0.5 m	20 kg	0.8 kg m s^2
Link 2	1 m	0.5 m	10 kg	0.2 kg m s^2

The Cartesian frame is fixed at the base of robot 1, and the trajectories of the object and the robots' end-effectors are specified relative to this frame. The task is to move the object forward and then backward in the x direction while keeping the height in the Y direction constant. The desired trajectory of the object is selected by a high-level planner as the reference input to the low level. If the two robots hold the object firmly, then the dynamics of the system are modelled as follows.

5.6.1   Dynamics of the object

Let $f_i = \begin{bmatrix} f_{ix}, & f_{iy} \end{bmatrix}^T$ be the force exerted by the end-effector of robot i on the object in Cartesian space. Then the motion of the object is described by

$$m\ddot{P} + mg = f, \qquad f = WF \equiv \begin{bmatrix} I_2, & I_2 \end{bmatrix} \begin{bmatrix} f_1 \\ f_2 \end{bmatrix} \qquad (5.14)$$

where m is the mass of the object, P is the position of the object in Cartesian space, g is the gravitational acceleration, f is the external force exerted on the object by the two robots, and I_2 is a 2×2 unit matrix.

---

[2]   The term **robot** will henceforth mean **robotic manipulator** and the two terms will be used interchangeably, unless stated otherwise.

From Eq.(5.14), one can see that, to achieve the object's specified acceleration, the combination of forces shared by the two robots is not unique.

5.6.2 Dynamics of each robot with servo controller

Suppose the two robots have an identical mechanical configuration, then the force-constrained dynamic equation of robot i in joint space is given by

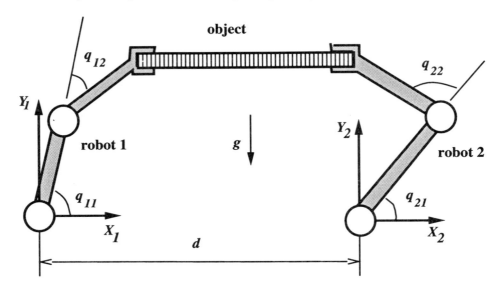

Figure 5.6 Two 2-link robots holding an object

$$H(q_i)\ddot{q}_i + C(q_i,\dot{q}_i)\dot{q}_i + G(q_i) + J_i^T f_i = \tau_i, \qquad i = 1, 2$$

where $q_i = [q_{i1}, q_{i2}]^T$ and $\tau_i = [\tau_{i1}, \tau_{i2}]^T$ are the vectors of the joint position and torque of robot i, respectively; $H(q_i)$ is the inertia matrix; $C(q_i,\dot{q}_i)\dot{q}_i$ represents the Coriolis and centrifugal forces; $G(q_i)$ represents the gravitational force; and J_i is the Jacobian matrix.

Suppose both robots are position-controlled with the computed torque algorithm. That is, the control input to robot i is

$$\tau_i = \hat{H}\left[\ddot{q}_{id} - K_{Di}(\dot{q}_i - \dot{q}_{id}) - K_{pi}(q_i - q_{id})\right] + \hat{h} \tag{5.15}$$

where $\hat{H}$ and $\hat{h}$ are the estimated values of H and $(C\dot{q}_i + G)$ respectively, q_{id} is the desired value of q_i, and K_{Di} and K_{pi} are the controllers' gains. The reference input to the system is the desired trajectory of the object specified by P_d, $\dot{P}_d$ and $\ddot{P}_d$, which will be transformed into the desired trajectories of the end-effector and the joints of each robot.

5.6.3 Problem statement

Suppose the object is a rigid body and there is no relative motion between the end-effectors and the object. For Eq. (5.14), let f_d and F_d be the desired values of f and F, respectively. Then we have

$$F_d = F_{Md} + F_{Id} \equiv W^* f_d + \left(I_4 - W^* W\right) y_0 \tag{5.16}$$

where $W^* \in R^{4 \times 2}$ is the pseudo-inverse of W, L_4 is a 4×4 unit matrix, and $y_0 \in R^4$ is an arbitrary vector in the null space of W. Therefore, the forces exerted by the end-effectors consist of two parts:

$$F_{Md} = \begin{bmatrix} F_{M1d} \\ F_{M2d} \end{bmatrix} \in R^4 \text{ is the force to move the object and}$$

$$F_{Id} = \begin{bmatrix} F_{I1d} \\ F_{I2d} \end{bmatrix} \in R^4 \text{ is the internal force.}$$

The following two problems arise: (1) sharing the moving force by the two robots, and (2) changing the internal force so as to satisfy a set of constraints, such as joint torque limits or energy capacity.

In Eq. (5.16), f_d can be specified by the desired trajectory. F_{Id} is given as the desired internal force, for example, $F_{Id} = 0$ for the least energy consumption. Because W^* is a constant matrix and both f_d and F_{Id} are specified, the desired force F_d is determined uniquely. However, this ideal situation of load sharing may not be achieved due to force and trajectory tracking errors. These errors may be caused by modelling/parameter errors, control performance tradeoff and/or disturbances. Therefore, it is necessary to share the load by, or to re-assign the load to, each robot dynamically. Our goal is to design a KBC to co-ordinate the two robots moving the object while minimising the internal force.

5.6.4 Principal output and its NN-based predictor

The reference inputs to the low-level subsystems are the desired acceleration $\ddot{P}_{id}$, velocity $\dot{P}_{id}$, and position P_{id} of robot i's end-effector, $i = 1, 2$. The internal force can be used to evaluate system performance and has an explicit relation to the reference inputs. Thus, the internal force is defined as the principal output. Because the force exerted by each robot to achieve a specified acceleration of the object is not unique, it is possible to adjust the internal force by modifying the reference inputs. Since the position tracking error needs to be kept small and the desired acceleration has an explicit relationship to the force exerted on the object, only the desired acceleration is modified in order to reduce the internal force. Then the desired acceleration issued to each robot is $\ddot{P}_{idm}$, this being the modified value of $\ddot{P}_{id}$, $i = 1, 2$. An NN-based predictor is designed to predict the force exerted on the object, which corresponds to each reference input. The predicted internal force (that is, the principal output) is then computed. The NN-based predictor has eight nodes at the INPUT layer, and the inputs are

$$P_{1d}(k), \qquad P_{1d}(k-1), \qquad P_{2d}(k), \qquad P_{2d}(k-1)$$

and $\qquad \ddot{P}_{1dm}(k), \qquad \ddot{P}_{1dm}(k-1), \qquad \ddot{P}_{2dm}(k), \qquad \ddot{P}_{2dm}(k-1)$

There are five HIDDEN nodes and six OUTPUT nodes with outputs:

$$\hat{f}_i(k+d/k), \qquad \text{for} \qquad i = 1, 2 \ , \quad d = 1, 2, 3$$

5.6.5 Simulation results

In the simulation, the task is for the two robots to move the object in the X direction from the initial position to the final position over a one-metre distance in five seconds, and then move it back to the initial position. The desired velocity and acceleration of the object are zero at both the initial and final positions. The sampling interval is $T_s = 0.01$ s. Force predictions are used for the modification process, and position tracking is achieved by the position controllers. The one-step ahead predictions $\hat{f}_i(k+1/k)$, $i = 1$, 2 are used in the KBC. The desired internal force is set to zero. The internal force error in the X direction without the KBC is plotted in Figure 5.7. After adding the KBC, the RMS error of the internal force in the X direction is reduced by 63%, as shown in Figure 5.8. Moreover, both the external force error and the position tracking error are kept almost the same as those without the KBC. Detailed results are summarised in Table 5.2. Since there is no motion in the Y direction, the internal force error in that direction is small enough not to require the KBC.

5.7 EXAMPLE 2: COLLISION AVOIDANCE FOR TWO 2-LINK ROBOTS

In this section, we present an example where the KBC co-ordinates two revolute robots for collision avoidance. A robot control system usually consists of a four-level hierarchy: task planning, path planning, trajectory planning and servo control. The problem of collision avoidance between robots can be solved at the path planning level by considering collisions between the robot and the fixed/static obstacles in the workspace. By path planning, we mean off-line geometric planning in the robots' workspace. Generally, there are two approaches to path planning: graph search and use of potential field [17]. Collision avoidance can also be achieved by finding collision-free trajectories using optimisation or search methods. In practice, the desired path and trajectory of each robot are determined by guiding the robot through the workspace with a joystick, and its servo controller is designed independently of, and separately from, the other robot's paths and trajectories.

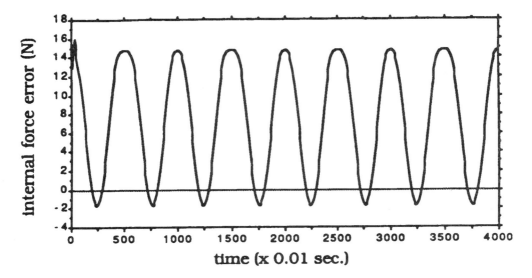

Figure 5.7 Internal force error in the *X* direction without the KBC

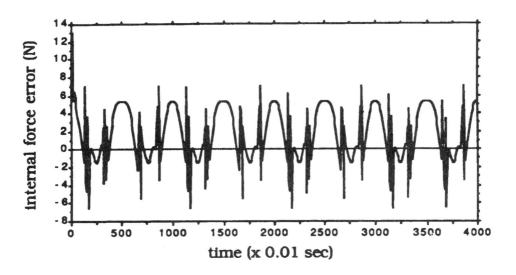

Figure 5.8 Internal force error in the *X* direction with the KBC

Regardless of the collision-avoidance scheme used, it is essential to track a robot's desired trajectory accurately, which in turn calls for high-performance servo controllers. Otherwise, collision may occur even if the desired trajectory is planned to be collision-free. This implies that the dynamics of multiple robots must be figured in their co-ordination. An on-line co-ordinator is thus needed to guide the robots using sensory information. This on-line co-ordination is commonly termed the path finding problem. Since path finding does not always guarantee that the robots achieve the goal, a high-level planner is still necessary. However, the existence of on-line co-ordination will ease the burden on both path planning and trajectory planning for collision detection and avoidance. The path finding problem for a multiple-robot system is the main subject of this section.

Table 5.2 The RMS errors of forces and position tracking

Sample intervals for statistics		RMS errors of internal forces (N)	
		without KBC	with KBC
0 – 1000	in X direction	9.58447	3.85020
	in Y direction	0.93141	0.53177
1001 – 2000	in X direction	9.57130	3.53340
	in Y direction	0.92339	0.49949
2001 – 3000	in X direction	9.57097	3.53688
	in Y direction	0.92339	0.49956
Sample intervals for statistics		RMS errors of external forces (N)	
		without KBC	with KBC
0 – 1000	in X direction	0.72359	0.95822
	in Y direction	2.54853	2.54345
1001 – 2000	in X direction	0.34883	0.70199
	in Y direction	0.01436	0.03345
2001 – 3000	in X direction	0.34883	0.70346
	in Y direction	0.01436	0.03355
Sample intervals for statistics		RMS tracking errors of object's positions (m)	
		without KBC	with KBC
0 – 1000	in X direction	0.03509	0.03529
	in Y direction	0.05733	0.05784
1001 – 2000	in X direction	0.03509	0.03530
	in Y direction	0.05759	0.05813
2001 – 3000	in X direction	0.03509	0.03530
	in Y direction	0.05759	0.05813

Most industrial robots are designed to work as stand-alone devices and are usually equipped with PID-type servo controllers. Thus, it is reasonable to make the following assumptions:

A1.   The desired path of a robot is obtained by teaching, and thus avoids collision only with fixed obstacles in the workspace.

A2.   Trajectory planning does not deal with the problem of avoiding collision between robots.

A3.   Collision avoidance is not a subject considered when designing servo controllers. Servo controllers are commercially designed and independent of one another.

A4.   No detailed knowledge on the dynamic structure and/or parameters of each robot and its servo controller is available.

It is in general very difficult to co-ordinate multiple robots under the above (realistic) assumptions. The problem of path finding is divided into two parts: (1) collision detection, and (2) collision avoidance. The former involves finding an algorithm and a set of rules for collision detection, and the latter involves the design of a manoeuvring strategy for collision avoidance.

5.7.1   Collision detection: the simplest case

The simplest configuration of multiple robots in a common workspace is shown in Figure 5.9 for two cylindrical robots. For simplicity, only collision avoidance in a two-dimensional workspace is considered, implying that there are no constraints on the vertical movement of the robots. The right-of-way is given to robot 1, that is, robot 1 will follow its desired trajectory, while robot 2 must be co-ordinated to avoid collision. Let $P_i(k)$ and $\phi_i(k)$, $i = 1$, 2, represent the position and angle of robot i at time k, respectively. Their d-step ahead predictions at time k are denoted by $\hat{P}_i(k+d/k)$, $\hat{\phi}_i(k+d/k)$. Then a fictitious, permanently colliding robot is defined with its position $P_c(k+d)$ and angle $\phi_c(k+d)$ as follows [16]:

$$P_c(k+d) = \sqrt{(d_{12})^2 + \left[\hat{P}_1(k+d/k)\right]^2 - 2d_{12}\,\hat{P}_1(k+d/k)\cos\left[\hat{\phi}_1(k+d/k) - \phi_{01}\right]} \quad (5.17)$$

$$\phi_c(k+d) = \ \arctan\ \frac{\hat{P}_1(k+d/k)\ \sin\left[\hat{\phi}_1(k+d/k)\right]}{d_{12} - \hat{P}_1(k+d/k)\ \cos\ \left[\hat{\phi}_1(k+d/k)\right]}$$

To guarantee collision avoidance in the presence of tracking error and prediction error, position and angular safety margins are defined by $P_s \geq 0$ and $\phi_s \geq 0$. Without loss of generality and also for simplicity, it is assumed that $\phi_{01} = 0$ and $\phi_{02} = \pi$. There are then six different possible configurations for the two robots. To detect a possible collision, the estimated angular margin is defined by

$$\Delta\hat{\phi}_s = \begin{cases} \Delta\hat{\phi} - \phi_s\,, & \text{if}\ \ \Delta\hat{\phi} \geq 0 \\ \Delta\hat{\phi} + \phi_s\,, & \text{if}\ \ \Delta\hat{\phi} < 0 \end{cases} \quad (5.18)$$

where

$$\Delta\hat{\phi} = \left[\phi_{02} - \hat{\phi}_2(k+d/k)\right] - \phi_c(k+d)$$

Then a set of rules can be derived for collision detection and one of them is listed below as an example.

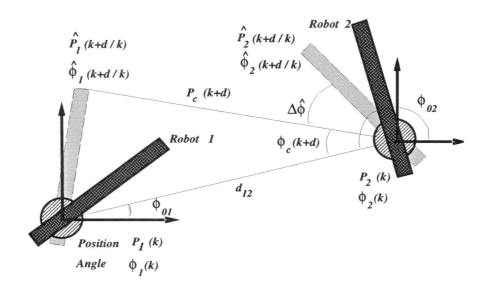

Figure 5.9 Configuration of two cylindrical robots in a common workspace

$$\text{IF } \phi_c(k+d) \geq 0$$
$$\text{IF } \Delta\hat{\phi}_s \leq 0 \text{ AND } \left|\Delta\hat{\phi}_s\right| \leq \left|\phi_c(k+d)\right| \text{ AND } \left[\hat{P}_2(k+d/k) \geq P_c(k+d) - P_s\right]$$
$$\text{THEN a collision is detected}$$

5.7.2 Collision detection for two revolute robots

For two revolute robots in a two-dimensional workspace, as shown in Figure 5.10, if $\Omega_0 \subset \Omega$ is the space in which the angle of link 2 is 0 or π for each robot, then, for each point $(x, y) \in \Omega / \Omega_0$, there are two different link configurations allowing the end-effector to reach the point. Robot 1 is designated to follow its desired trajectory, while robot 2 has to be co-ordinated for collision avoidance. Under assumptions A1 to A4, this redundant case is more difficult to co-ordinate than that of cylindrical robots, as shown in Figure 5. 9.

Let $q_{i1}(k)$, $q_{i2}(k)$ be the joint angles of robot i at time k, $i = 1, 2$, and let $\hat{q}_{i1}(k+d/k)$ and $\hat{q}_{i2}(k+d/k)$ be their d-step ahead predictions, respectively.[3] Two fictitious permanent colliding robots, corresponding to the two possible configurations of robot 1, can be defined with positions P_{c0}, P_{c1} and angles ϕ_{c0}, ϕ_{c1}, as shown in Figure 5.11.

---

[3]   In what follows, $\hat{q}_{i1}(k+d/k)$ and $\hat{q}_{i2}(k+d/k)$ will be represented by $\hat{q}_{i1}$ and $\hat{q}_{i2}$, respectively, to simplify the notation.

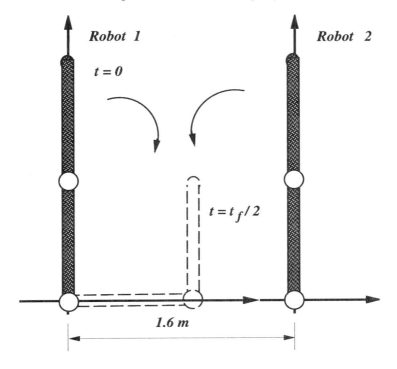

Figure 5.10 Configuration of two robots in a common workspace

$$P_{c0} = \sqrt{(\hat{P}_{10})^2 + (d_{12})^2 - 2\hat{P}_{10}d_{12}\,\cos\hat{\phi}_{10}}, \quad \text{and} \quad \phi_{c0} = \arctan\left(\frac{\hat{P}_{10}\,\sin\hat{\phi}_{10}}{d_{12} - \cos\hat{\phi}_{10}}\right) \quad (5.19)$$

$$P_{c1} = \sqrt{(L_{11})^2 + (d_{12})^2 - 2L_{11}d_{12}\,\cos\hat{q}_{11}}, \quad \text{and} \quad \phi_{c1} = \arctan\left(\frac{L_{11}\,\sin\hat{q}_{11}}{d_{12} - \cos\hat{q}_{11}}\right) \quad (5.20)$$

where L_{ij} is the length of link j of robot i, d_{12} is the distance between the bases of the two robots, and

$$\hat{P}_{i0} = \sqrt{(L_{i1})^2 + (L_{i2})^2 + 2L_{i1}L_{i2}\,\cos\hat{q}_{i2}}, \quad i = 1, 2,$$

$$\hat{\phi}_{i0} = \arcsin\left[\frac{L_{i1}\,\sin\hat{q}_{i1} + L_{i2}\,\sin(\hat{q}_{i1} + \hat{q}_{i2})}{\hat{P}_{i0}}\right], \quad j = 1, 2.$$

We want to define a permanent colliding robot corresponding to different values of P_{c0}, P_{c1} and ϕ_{c0}, ϕ_{c1}. The position of the permanent colliding robot can be conservatively selected as

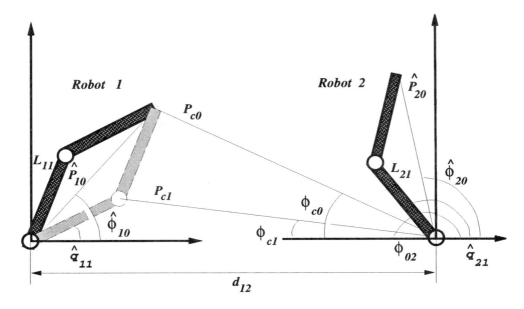

Figure 5.11 Collision detection of two revolute robots

$$P_c = \min \ (P_{c0}, \ P_{c1}) \tag{5.21}$$

and its angle ϕ_c will be defined according to different values of ϕ_{c0} *and* ϕ_{c1}.

Then a set of rules can be derived to detect collisions. As an example, in our simulation experiments, the desired trajectories of the robots in Figure 5.10 are symmetric. Some special points of the desired trajectory are given in the Table 5.3, where the subscript d represents the desired value. Since the moving directions of both robots are the same in this example, one can derive a suitable permanent colliding robot by choosing

$$\phi_c = \max \ (\phi_{c0}, \ \phi_{c1}) \qquad \text{and} \qquad P_c = \min \ (P_{c0}, \ P_{c1})$$

Then the rules for collision detection can be simplified and are listed below:

1.   Compute $\hat{\phi}_2 = \max \ (\hat{\phi}_{20}, \ \hat{q}_{c1})$ and $\hat{P}_2 = \max \ (\hat{P}_{20}, \ L_{21})$

2.   Compute angular and length margins $\Delta\hat{\phi}_s$ and $\Delta\hat{P}_s$

$$\Delta\hat{\phi}_s = \begin{cases} \Delta\hat{\phi} - \phi_s, & \text{if } \Delta\hat{\phi} \geq 0 \\ \Delta\hat{\phi} + \phi_s, & \text{if } \Delta\hat{\phi} < 0 \end{cases} \qquad \text{and } \ \Delta\hat{P}_s = P_c - \left(\hat{P}_2 + P_s\right) \tag{5.22}$$

where $\Delta\hat{\phi} = \left(\phi_{02} - \hat{\phi}_2\right) - \hat{\phi}_c$

3.   Detect collision using the following rules:

R1: IF $\phi_c \geq 0$ AND $\Delta\hat{\phi}_s \leq 0$ AND $\left|\Delta\hat{\phi}_s\right| \leq \left|\phi_c\right|$ AND $\Delta\hat{P}_s \leq 0$

THEN a collision is detected.

R2: IF $\phi_c < 0$ AND $\Delta\hat{\phi}_s \geq 0$ AND $\left|\Delta\hat{\phi}_s\right| \leq \left|\phi_c\right|$ AND $\Delta\hat{P}_s \leq 0$

THEN a collision is detected.

Table 5.3 The desired trajectories in simulation

Robot 1	Joint 1			Joint 2		
	q_{11d}	$\dot{q}_{11d}$	$\ddot{q}_{11d}$	q_{12d}	$\dot{q}_{12d}$	$\ddot{q}_{12d}$
initial values, $t = t_0$	90°	0	0	0°	0	0
middle point, $t = t_f/2$	0°	max	max	90°	0	0
final values, $t = t_f$	−90°	0	0	0°	0	0
Robot 2	Joint 1			Joint 2		
	q_{21d}	$\dot{q}_{21d}$	$\ddot{q}_{21d}$	q_{22d}	$\dot{q}_{22d}$	$\ddot{q}_{22d}$
initial values, $t = t_0$	90°	0	0	0°	0	0
middle point, $t = t_f/2$	180°	max	max	−90°	0	0
final values, $t = t_f$	270°	0	0	0°	0	0

5.7.3 Strategy for collision avoidance

Once a collision is detected, the desired trajectory of robot 2 will be modified in order to avoid the collision. Though either increasing or decreasing the speed of robot 2 may avoid the collision, the reasonable manoeuvre is to slow down robot 2 since the maximum speed and acceleration/deceleration are usually bounded. This implies that the reference input be modified in one direction. In order to give robot 2 a sufficient time so that it can manoeuvre to avoid the anticipated collision, the safety margins ϕ_s and P_s are added to the length and angular margins as in Eq. (5.22). Moreover, it is possible to modify the desired trajectory of one joint to avoid collision, which will in turn simplify the modification process of the reference inputs.

5.7.4 Simulation results

The dynamic and kinematic parameters of the simulated robots are identical as listed in Table 5.1. It is assumed that each robot is equipped with a PD-type servo controller, and the sampling interval is 0.01 s. An NN-based predictor is designed for each robot. There are six INPUT nodes with inputs

$$q_{i1d}\ (k),\ q_{i1d}\ (k-1),\ q_{i1d}\ (k-2),\ q_{i2d}\ (k),\ q_{i2d}\ (k-1), q_{i2d}\ (k-2)$$

eight HIDDEN nodes and six OUTPUT nodes with outputs,

$$\hat{q}_{i1}(k+d\,/\,k)\,,\ \hat{q}_{i2}(k+d\,/\,k),\qquad d=1,\ 2,\ 3,\ \text{for robot}\ i:\ i=1,\ 2\,.$$

The angular and length safety margins are set to $\phi_s = 0.2$ rad and $P_s = 0.05$ m, and two-step ahead predictions are used for collision detection. Figure 5.12 shows the actual trajectory of robot 2, showing the slow-down of robot 2 to avoid collision.

5.8 CONCLUSION

Focusing on the problem of co-ordinating multiple systems, a knowledge-based co-ordinator is designed using the techniques of both intelligent control and neural networks. As the high-level co-ordinator in a hierarchical structure, its basic principle is to modify the reference inputs of low-level subsystems according to the principal output prediction in order to achieve the desired performance. By adding the proposed KBC, the internal structure and parameters of the low-level subsystems are not affected. Hence, each servo controller of the low-level subsystems can be designed separately from, and

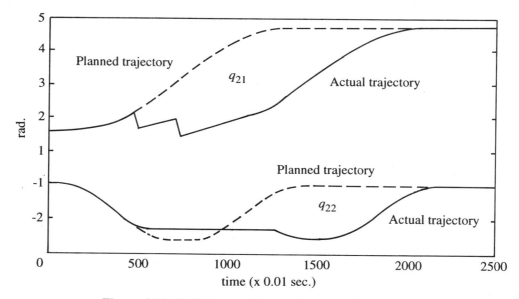

Figure 5.12 Collision avoidance of two revolute robots

independently of, the others; no constraints need to be imposed on the design of low-level controllers. This implies that some commercially designed servo controllers for a single system can be co-ordinated to work for multiple systems.

Using the principal output and its prediction, and the structure of the decision tree for knowledge representation, the knowledge base necessary to co-ordinate multiple systems is greatly simplified while guaranteeing system stability. By using a predictor, the negative effect of system time delay is eliminated and each reference input is analysed before putting it in operation. The unknown parameters and/or time-varying properties of a multiple system are handled by the NN-based predictor, while leaving the logical reasoning and decision making on co-ordination to the KBC.

To test this new scheme, the co-ordination problem for two 2-link robots holding a rigid object was simulated. By modifying the reference input of each robot, the internal force exerted on the object was reduced by 63%. The proposed scheme was also tested for a loosely coupled system, i.e collision avoidance for two revolute robots. It is assumed that both path planning and trajectory planning did not consider collision avoidance, and adding the KBC did not impose any constraints on the design of the servo controller. This may relax the usual requirements (for example, the knowledge of exact dynamics) imposed on path planning, trajectory planning and servo controllers design. The simple structure and algorithm, no constraints on the design of individual servo control systems, and good simulation results indicate the scheme's potential for the effective co-ordination of multiple systems.

REFERENCES

[1]   R. E. Larson, P. L. McEntire and J. G. O'Reilly [Eds.] (1982), *Tutorial: distributed control* (2nd edition), IEEE Computer Society.

[2]   S. Lee and M. H. Kim (1987), Cognitive control of dynamic systems, *Proc. of the 2nd IEEE Int. Symp. on Intelligent Control*, 455-460.

[3]   Z. Geng and M. Jamshidi (1988), Expert self-learning controller for robot manipulator, *Proc. IEEE Int. Conf. on Decision and Control*, 1090-1095.

[4]   A. K. Krijgsman, H. M. T. Broeders, H. B. Verbruggen and P. M. Bruijn (1988), Knowledge-based control, *Proc. IEEE Int. Conf. on Decision and Control*, 570-574.

[5]   J. Jiang and R. Doraiswami (1987), Information acquisition in expert control system design using adaptive filters, *Proc. of IEEE 2nd Int. Symp. on Intelligent Control*, 165-170.

[6]   K. L. Anderson, G. L. Blankenship and L. G. Lebow (1988), A rule-based adaptive PID controller, *Proc. IEEE Int. Conf. on Decision and Control*, 564-569.

[7]   B. Porter, A. H. Jones and C. B. McKeown (1987), Real-time expert controller for plants with actuator non-linearities, *Proc. of IEEE 2nd Int. Symp. on Intelligent Control*, 171-177.

[8]   G. K. H. Pang (1988), A blackboard control architecture for real-time control, *Proc. America Control Conference*, 221-226.

[9]   K. G. Shin and X. Cui (1991), Design of a knowledge-based controller for intelligent control systems, *IEEE Trans. on Systems, Man and Cybernetics*, **32**(2), 368-375.

[10] G. Cybenko (1989), Approximation by superpositions of a sigmoidal function, *Mathematics of Control, Signals and Systems*, **2**(4), 303-314.

[11] V. Vemuri [Reprint edited by] (1988), *Artificial neural networks: theoretical concepts*, IEEE Computer Society.

[12] P. J. Werbos (1988), Back propagation: past and future, *Proc. Int. Conf. on Neural Networks*, **1**, 1343-1353.

[13] D. E. Rumelhart, and J. L. McCelland (1986), *Learning internal representations by error propagation, parallel distributed processing: explorations in the microstructure of cognition, Vol.1: Foundations*, MIT Press.

[14] K. G. Shin and X. Cui (1991), Design of a general-purpose MIMO predictor with neural networks, *Proc. of the 13th IMACS World Congress on Computation and Applied Mathematics*, Dublin, Ireland.

[15] H. Asada and J. J. E. Slotine (1986), *Robot analysis and control*, Wiley-Interscience.

[16] E. Freund and H. Hoyer (1988), Real-time pathfinding in multirobot systems including obstacle avoidance, *Int. J. of Robotics Research*, **7**(1), 42-70.

[17] O. Khatib (1986), Real time obstacle avoidance for manipulator and mobile robots, *Int. J. of Robotics Research*, **5**(1), 90-98.

ACKNOWLEDGEMENT

The work reported in this chapter was supported in part by the National Science Foundation under grant number IRI-9209031. Any opinions, findings and recommendations expressed in this publication are those of the authors and do not necessarily reflect the view of the NSF.

6

Neural networks for mobile robot piloting control

L. Moreno[1], M.A. Salichs[2], D. Gachet[2], J. Pimentel[3], F. Arroyo[1] and A. Gonzalo[1]
[1] Universidad Politécnica de Madrid, Dpto Ingeniería de Sistemas y Automática, Spain
[2] Universidad Carlos III, Escuela Politécnica Superior, Madrid, Spain
[3] General Motors Institute, Flint, Michigan, USA

6.1 INTRODUCTION

The process of piloting autonomous mobile robots in unstructured, dynamic, and complex environments is difficult. A multitude of problems arise such as noisy measurements, lack of appropriate sensors, difficulty in achieving real-time sensor processing, and constructing an appropriate model of the environment based on sensory information. For example, when working with ultrasonic range sensors there are inaccuracies in the values measured due to noise or to the physics of the ultrasound waves. Thus obtaining an accurate state of the environment is difficult.

In order to avoid these problems, the UPM mobile robotics research group has developed a reactive piloting architecture, named AFREB (Adaptive Fusion of Reactive Behaviours). This architecture follows a behavioral methodology where all robot tasks are performed in terms of a set of elementary tasks referred to as primitive behaviors. The fundamental idea of behavioral control is to view a robot mission or task, also referred to as an emergent behavior as the temporal execution of a set of primitive behaviors. One of the most important problems in behavioral control involves the determination of the appropriate mixture of primitive behaviors in order to execute complex tasks.

There are many examples of behavioral navigation. One of them is the Brooks subsumption architecture [4] that decomposes the architecture into behaviors. One of the main difficulties of this aproach lies in the control of behaviors. It has to be decided which of the behaviors should be active and obtain control over the actuators at a particular time. This problem has been solved in different ways: hand precompilation of the control flow and priorities [4] and automatic recompilation by using a description of the desired behavior selection [10]. In both cases there is a completely fixed modelization of the relationship and switching between behaviors. These interactions are defined by means of the subsumption language and each substrate is organized into a series of incremental layers built by a network of finite state machines with timing elements.

Payton has considered the need for integrating high-level planning activities with lower-level reactive behaviors [18,19]. The internalized plans can be thought of as representations that allow the raw results of search in any abstract state space to be made available for direct use within continuous real-time decision-making processes.

The reactive navigation proposed by Arkin for holonomous vehicles [2,3] doesn't use arbitration for coordinating the multiple active agents; the behaviors are modelled by using a potential field formulation to describe the reactions of the robot to the world. The outputs of each individual behavior reacting to sensory information coming from the environment comprise a velocity vector representing the direction and the speed at which the robot is to move given current environmental conditions. Each of these motor schemas is computed asynchronously for the point where the robot is currently located. The perceptual schemas are embedded within motor schemas providing the information that is required for them to compute their reaction to the world. No abstract model of the world is built to reason over during plan execution.

Maes and Brooks [16] have reported a scheme where a legged robot learns how to walk using a behavioral methodology. Each behavior tries to learn when it should become active. Learning is based on positive and negative feedback functions and takes into account relevance and reliability. Negative feedback is received by all of the behaviors every time at least one of the touch sensors fires. Positive feedback is received every time the wheel measures forward movement. It is difficult to apply this method to situations involving a complex state (e.g., mobile robots navigating in a cluttered dynamic environment). This approach is a highly intelligent, distributed system for applications where it is relatively straightforward to determine if the system is making progress to a goal. The method cannot be easily generalized to other applications.

Clark, Arkin, and Ram [6] have described a system that adapts its behavior based on recent experience, taking into account robot performance and changes in the environment. They argue that if something works well, keep doing it (perhaps harder) but if things are not proceeding well then try something different (the system builds momentum). Behavioral contributions are changed continuously and gradually using a rule-based approach. The system does not really learn, as the performance measures are used only to define four situations. A rule-based system is then used to define a table of situation-action. This is an example of rote learning, where the knowledge injected into the system uses a set of performance measures to find out in what situation the robot is located.

Salganicoff and Bajcsy [27], have reported an experiment which consists of learning to position a gripper based on visual observation. The learning is empirical in nature, and is done by having the robot observe repeated interactions with the task environment. Several servorimotor primitives are implemented. The working hypothesis is that inductive learning must happen incrementally with respect to the number of parameters to be characterized, otherwise the learning becomes intractable due to the combinatorics of the task which requires an unrealistically large number of samples. A similar experiment was reported by van der Smagt, Krose, and Groen [30].

Cheng, Xiao, and Le Quoc [5] have used a back-propagation NN to control an AGV. The input layer consists of two neurons representing the tracking errors from the camera image processor, and the sole neuron in the output layer provides the command signal for the raw rate.

Shibata *et al.* [29] have described a hybrid symbolic and neuromorphic control for autonomous robots. The system develops a control strategy by symbolic reasoning and adapts to the environment by neuromorphic control. A hierarchical intelligent control system is used consisting primarily of two levels: an adaptation level and a learning level. The adaptation level is basically a feedback control loop with a neural servo-controller. The learning level contains the symbolic controller, a data base, and the knowledge base.

Lewis and Bekey [13] describe a system of nanorobots which cooperate in a synergistic fashion to achieve a common goal (that of destroying malignant tissue), where each robot acting alone has a low probability of achieving its purpose. Kweon *et al.* [11] have developed a robust system using a behavioral methodology similar to ours but they do not report any experiments involving learning. Rao *et al.* [23] discuss the N-learner problem where the learners are each capable of learning concepts and the performance of the composite system constituted by the fuser and the individual learners is studied in a theoretical fashion.

6.2   AFREB: AN ARCHITECTURE FOR BEHAVIORAL CONTROL

Several techniques can be used for the appropriate simultaneous activation of primitive behaviors to execute more complex robot behaviors. These techniques have been implemented within an architecture which allows the execution of various behaviors either simultaneously or sequentially. The architecture follows a multi-behavioral approach in which a set of behaviors is defined, each of which causes the robot to respond in a specific manner to detected stimuli [1-2].

Several researchers have already argued the importance of looking at a mobile robot as a set of primitive behaviors. Primitive behaviors are also important components of reactive control, which is a recently emerged paradigm for guiding robots in unstructured and dynamic environments [16,1,2,4,10]. Mobile robots must continuously interact with their environment and this is the essential characteristic of reactive programs. By **reactive** is meant that all decisions are based on the currently perceived sensory information. An important consequence of the reactive approach is that for the piloting level we don´t need any environmental modelling. As depicted in Figure 6.1, the fundamental idea of behavioral control is to view a robot task (also referred to as an emergent behavior) as the simultaneous and temporal execution of a set of primitive behaviors. We define an emergent behavior as a simple or complex task or a task which is made up of more elementary (primitive) behaviors. Whereas primitive behaviors   correspond   to   robot

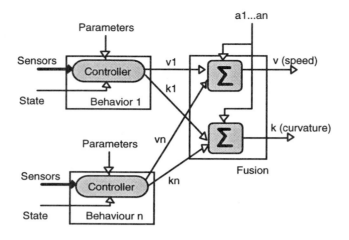

Figure 6.1 Basic idea of behavioral control

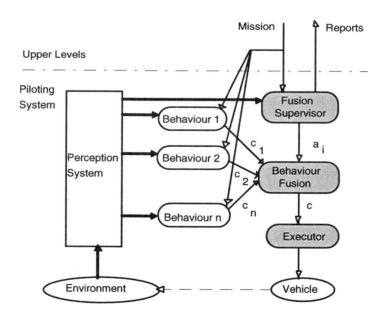

Figure 6.2 AFREB architecture

actions, emergent behaviors correspond to missions, goals, or subgoals. The method implemented is reactive, behavior-based, and with simultaneous activation of behaviors [21].

A behavioral control architecture called Adaptative Fusion of REactive Behaviors has been developed for experimenting with a wide range of reactive control methodologies [7,8]. The AFREB architecture is illustrated in Figure 6.2. and basically consists of the following modules: behavior fusion module, fusion supervisor, behavior primitives, and executor. The main function of the fusion supervision module is the calculation of the weights of each primitive behavior. The function of the behavior fusion module is the calculation of the actual robot commands based on the primitive behavior weights generated by the fusion supervision module. The function of the executor function is the execution of the actual robot commands while enforcing control limitations due to non-holonomic constraints and the specific characteristics of each vehicle.

We have used the behavioral architecture in several simulation and actual experiments. All simulation work has been performed with OPMOR, a simulator which is a constituent component of the software development environment [20]. Results with the actual robot correspond to the UPM Robuter platform [24].

The inputs to the executor are speed v and curvature κ. A primitive behavior can be characterized by a temporal sequence of appropriate values for v and κ which cause the robot to exhibit the pre-specified primitive behavior. Thus we define the output of a primitive behavior $c(k)$ as the vector $c(k)=(v,\kappa)$ where the variable k denotes the kth cycle of the robot controller. In what follows we will drop the index k for notational simplicity.

If we denote $c_1, c_2, ..., c_n$ as the output of each primitive behavior, then the output of an emergent behavior (i.e., mission, task, or subtask) can be obtained as a linear combination of the primitive behavior outputs, that is:

$$c_0 = \frac{\sum_{i=1}^{n} a_i \cdot c_i}{\sum_{i=1}^{n} a_i} \qquad (6.1)$$

with

$$\sum_{i=1}^{n} a_i \neq 0 \qquad (6.2)$$

where the a_i coefficients, with $0 \leq a_i \leq 1$ are found by an appropiate combination of measurement information provided by the perception system. Other fusion methods to obtain the emergent behavior are also feasible, but this method is simple and efficient. As noted, one of the most challenging problems in behavioral control involves the learning of the appropriate mixture of primitive behaviors in order to execute complex tasks. These a_i coefficients are dynamically changed. Thus, the main function of an intelligent controller is the learning of the weights a_i so that the performance of the robot for the execution of the tasks is adequate. The primary goal of the following sections is the discussion of techniques to generate the appropriate values of the coefficients a_i based on the task to be performed and based on the perception system. There are two possibilities for learning the coefficients a_i: on line or off line. For autonomous systems the on line learning method is the most interesting. Whether or not learning is present it is to be noted that the coefficients a_i are dynamically adjusted by the controller when the robot is in motion.

As noted, the outputs of the primitive behaviors are fused in a linear combination fashion to produce emergent behaviors. The set of primitive behaviors active at any given time depends upon the occurrence of specific events in the environment (e.g., the detection of an obstacle, time for the robot to recharge itself, the issue of an interactive command, etc.). The primitive behaviors (i.e., actions) which have been implemented are:

C1. goal attraction
C2. perimeter following (CW, clockwise)
C3. perimeter following (CCW, counterclockwise)
C4. free space
C5. keep away (from objects)
C6. follow a predefined path.

The goal attraction primitive produces an output which directs the robot towards a specific goal. The perimeter following behavior follows the perimeter of a fixed (i.e., static) obstacle, keeping a prespecified distance away from it. This behavior takes into account the minimum radius of curvature $R = v/w$ of the mobile robot in order to avoid uneven corners. The movement can be performed clockwise (CW) or counterclockwise (CCW). An example of this behavior is shown in Figure 6.3.

The free space behavior will cause movements in a direction such that the robot frees itself from objects. This behavior takes the information from measurements provided by the front sensors and chooses the direction of the longer adjacent measures in order to ensure a safe course in that direction. The output of this behavior is the direction of motion at a constant speed.

The keep away behavior basically avoids obstacles in the proximity of the robot, choosing directions that ensure no collision with nearby obstacles. The remaining behaviors are self-explanatory.

There are several reasons why defining an emergent behavior as a linear combination of primitive behaviors is a good idea. The first reason is that it appears that humans perform actions in a similar way. For example, when we walk towards a goal and at the same time we want to be away from specific objects, it appears that we move simultaneously in a direction that is a combination of the direction of the goal and that which is away from the objects. The same observation can be made for the velocity of the movement. A second reason is that it allows the control of the robot navigational task at a higher level in terms of the primitive behaviors instead of the lower-level velocity and curvature commands. A final reason is that the linear combination includes the switched case when all the coefficients a_i are binary (i.e., 1 or 0).

Figure 6.3 Example of contour following behavior

Likewise, we have implemented the following emergent behaviors (i.e., tasks or missions):

- M1, corridor navigation,
- M2, navigation following a predefined path,
- M3, navigation between two points (i.e., towards a goal),
- M4, surveillance of the environment.

In general, a mission is executed while taking into account the dynamic nature of the environment as sensed by the perception system of the robot. For example, mission M2 can be performed in terms of only primitive behavior C6 as long as there are no objects (static or moving) blocking the path. To account for the dynamic nature of the environment, we have implemented M2 in terms of C2 through C6.

Some missions require a small set of primitive behaviors whereas other missions require a larger set. The system handles addition or removal of primitive behaviors in an easy manner. One thing to keep in mind in the process of adding or removing primitive behaviors is that the set of currently defined primitive behaviors depends on the mission.

6.3 BEHAVIOR FUSION SUPERVISOR

The main problem can be formulated as follows. How can the system learn or how can the system determine the appropriate values for the a_i coefficients (i.e., mixture of primitive behaviors) in order to execute a given task Mi? In the context of the architecture discussed in the previous section, the learning process is performed within the fusion supervisor module. The behavior fusion module calculates the final robot commands based on the perception system in a dynamic fashion. Thus while the robot is in motion, one can view the a_i coefficients as being dynamic in nature.

6.3.1 Learning in autonomous mobile robots
The type of learning method applicable to a mobile robot depends on the type of information which the mobile robot has to learn. These robot learning approaches can be divided into four categories:

- *Learning for calibration or parameter adjustments.* This type of learning tunes the existing controllers or programs.
- *Learning about the world.* In this type of learning some representation of the world is constructed, updated, and refined, based on information acquired from the sensor system.
- *Learning to coordinate behaviors.* This type of learning uses existing behavior primitives which have to be coordinated to produce adequate reactions to external situations. Basically, coordinating behaviors attempts to determine when behaviors are to be executed.
- *Learning new behaviors.* This type of learning requires the building of new sequences from primitives actions.

The problem in our case is learning to coordinate the primitive behaviors to achieve a determined mission. This coordination is made by the behavior fusion supervisor, by a proper selection of the primitive behavior coefficients. To cope with the learning problem, it is possible to follow different approaches: expert systems, fuzzy systems, heuristics, neural networks, or genetic algorithms.

All learning approaches to be discussed next have basically the same objective: that of learning the control rules to execute a given mission given a set of examples, advice, or solutions to similar problems. Within the context of sensor-based autonomous systems, generating control rules typically involves obtaining environmental information through a perception system, identifying a set of situations, and generating the appropriate control actions based on the state and performance of the system.

Although it is currently difficult to provide a good classification of learning methodologies because of the various criteria and contexts used, in the following we provide a classification which is appropriate for autonomous mobile robots.

- *Rote learning.* Rote learning is synonymous with explicit programming and is analogous to memorization in humans. Under this scheme, learning is actually performed external to the system (usually by a human being) and injected into the system once all the learning has been completed. A crucial step in this scheme is the representation of knowledge in some suitable format. Once the knowledge is in the system, the system does not learn further (i.e., it stops learning). The rote learning approach basically consists of a set of situation-action rules, taking into account the performance level where the set of rules is generated off line by a human.

Rote learning simply involves introducing the knowledge directly into the system via a computer program or a database. There are many learning examples in this category as it is the simplest one to implement. Two well-known paradigms (from the viewpoint of learning) in the rote learning category are an expert system and a fuzzy system where the knowledge is introduced as a computer program in a suitable format. The primary distinction between an expert system and a fuzzy system is that the knowledge base and the processing logic is symbolic in the former and fuzzy in the latter. A third paradigm involves heuristics where the control rules are generated in an *ad-hoc* manner.

- *Advice taking.* This is also known as learning from instruction or learning by being told. For example, given control information in the form of heuristics, a new set of rules (for example of the form if ... then ...) is generated. The system is required to transform the knowledge of the entity giving advice into an appropriate form to be of effective use and will incrementally augment the system's knowledge. Learning from taking advice basically involves translating the advice into a set of control rules. No example is known for autonomous systems.

- *Learning from examples or evidence.* This is a common and powerful learning paradigm that may be implemented in a variety of ways. Given a set of examples and counterexamples of control actions, the system induces actions that will hopefully include all of the positive examples and none of the counterexamples. An advantage of this system is that it does not require representation of knowledge in any way. The source generating the examples may include a teacher, the system itself, or the external environment. Thus this system generates a set of control rules autonomously given a set of human-generated examples. The examples available may include only positive examples or both positive and negative examples. Many NN architectures, e.g. back-propagation, are good examples of learning from examples.

- *Learning by exploration or observation.* Nature provides the best example of this paradigm for learning through the generation of some actions and consequent survival of beings. Learning by observation does not include a teacher and consists of a number of processes such as:

  (i) creating classifications of given observations
  (ii) discovering relationships and laws governing a process

(iii) forming a theory to explain a given phenomenon.

A learning system that is built based on this technique is not provided with any sets of instances or examples of a concept. In addition, the learning system may have to deal with a number of observations that represent several concepts rather than just concentrating on a single concept at a time. In the learning approach based on exploration or observation, the system generates both a set of control rules and appropriate examples autonomously. The NN learning approach of reinforcement is a good example of learning by exploration or observation.

- *Learning by analogy.* This involves the solution to a new problem by adapting a known solution to a similar problem. The system generates a set of control rules based on a solution provided by a human. No examples are known for autonomous systems.

With the exception of rote learning, the remaining learning approaches learn the control rules autonomously. The learning approach from examples or evidence requires that the examples be provided explicitly by a human (or teacher), whereas the learning approach by exploration or observation generates its own examples autonomously.

Two different groups of solutions based on different learning methods have been adopted to solve the behavior supervision problem:

1. Rule-based supervisors, where a programmer includes his experience with the system in a certain number of rules.
2. Neural-based supervisors, where the system admits three learning schemes which involve: a) learning from examples, b) integrating expertise information and learning from examples, and c) learning from exploration.

All of the schemes discussed in the following sections have been initially developed and simulated in OPMOR and then implemented and debugged on the Robuter platform, see Figure 6.4, based on the architecture described previously. More specifically, each scheme corresponds to a specific fusion supervisor module. The other modules of the architecture remain unchanged.

For the mobile robot Robuter used in our laboratory, the environment information is given by a set of twenty-four ultrasonic range measurements and the location of the vehicle is provided by odometry reading.

6.3.2 Neural network based supervisors
The incorporation of learning in autonomous mobile robots offers the following advantages:

- Learning the appropriate control actions for a given situation is often more efficient than programming the knowledge directly into the robot (i.e., rote

Figure 6.4 Platform used in actual experiments

learning). Thus the problems of knowledge representation, acquisition, and programming by a human are obviated or simplified. In other cases, the equivalent program would be too complex.

- Learning allows the discovery of control actions not foreseen by humans. The size of the space of all experiences is huge and depends on the problem domain. Thus finding the best control action for each situation is sometimes impossible. In some cases it is easier to define a learning algorithm than to define and express the knowledge required to generate appropriate control actions. In this case a human being can actually learn or discover new facts from the autonomous system.
- Learning is crucial to achieving a truly autonomous intelligent system. As noted, most definitions of intelligent systems incorporate learning as a constituent element.

- Learning allows the discovery of control actions by observation, experimentation, and discovery. The control actions can be a set of rules. Appropriate values of parameters can be also found (identification). The old saying 'one learns from one's mistakes' is applied here. The more mistakes are made, the more the system learns.
- Learning allows the system to identify unexpected events, to self-monitor, and to find causes for events.
- Learning allows systems to improve their performance and extend their functionality.
- Learning helps to automate the process of knowledge acquisition.
- Learning is important as a research area, as new and improved methods can be found. For instance, the limitations of current approaches can be found and solutions devised [25]. Also, by using current learning paradigms to characterize the problems, they can be improved.

6.3.2.1 Feedforward
The first attempt to implement a behavior fusion supervisor module, which learns from examples the successful completion of mission M3 (collision free navigation between the start and goal positions), was made with a feedforward multilayer network and using the backpropagation learning rule. The supervisor structure is shown in Figure 6.5.

The training process belongs to the category of learning from examples where a teacher is used to generate the actions. The neural network system learns the appropriate simultaneous activation of primitive behaviors in order to execute more complex robot behaviors (missions or tasks). The system uses a supervised learning technique with a human trainer generating appropriate training samples for the simultaneous activation of behaviors in a simulated environment using OPMOR [20]. The method exploits the fact that humans are adept at assessing complex situations based on qualitative data, but their goal functions are difficult if not impossible to capture analytically. A major advantage of neural networks is that they can imitate behavior based on these elusive rules without formally specifying them. Humans have the best perception system of any machine ever devised in terms of vision and detection of special conditions; and so far humans are the best strategists.

The NN is trained by a human by means of an OPMOR control window, providing control for each behavior independently, as shown in Figure 6.6. The values are changed using a slider panel of X Windows in an on line fashion when the simulator is running and read by the simulator to control the robot.

The feedforward network has an input layer of fifteen nodes, two hidden layers with thirty nodes each, and an output layer with five nodes. The fifteen inputs correspond to eleven ultrasonic range measurements, the current heading of the robot (relative to the goal position), its current velocity, the remaining Euclidean distance to the goal, and an additional input insight used for indicating whether the robot can move in the direction of the goal. The five outputs correspond to the a_i weights associated with the five primitive behaviors used in this mission.

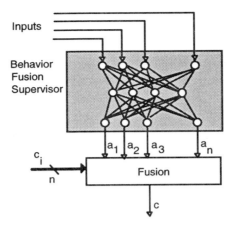

Figure 6.5 Fusion supervisor implemented with a feedforward NN.

The process of learning in this method attempts to find a relationship between a set of sensor measurements and a set of behavior coefficients at each control cycle. The error between the desired and the actual outputs of the neural network is used to train the network by the backpropagation algorithm. The performance levels are taken into account implicitly by defining the current heading of the robot (relative to the goal position) and the remaining Euclidean distance to the goal as inputs to the network.

One advantage of this method is that it avoids conventional programming while learning to perform the task and it does not require a knowledge representation. The control rules are not specified explicitly, as the system generates them in a natural way with respect to the problem domain.

In spite of the ability to learn from the examples presented to the network, there are some important difficulties in training the network with a human trainer. One of the difficulties originates in the method used to fuse the primitive behavior. The fusion process is done as a linear combination of the primitive behaviors' output, and for this reason there are multiple combinations of primitives able to provide an acceptable output. Another, and probably the most important, difficulty is due to the fact that the human trainer can see the whole environment instead of a fragmentary image provided by the eleven ultrasound sensors. Besides, the training window provides a slider to modify each of the weights associated with the primitive behaviors, and the human trainer can only modify one of the sliders at a given moment, which causes errors and delays in adjusting the weights.

Figure 6.7 illustrates the performance of the NN in a real environment different from the ones used for training.

In order to solve some of these problems, this architecture was modified to decouple the output weights from the situation identification.

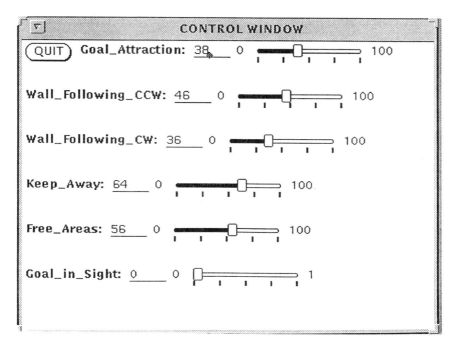

Figure 6.6 Training window

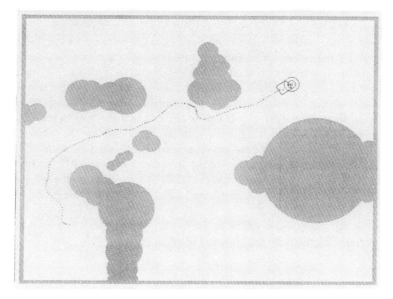

Figure 6.7 Performance of the feedforward NN supervisor

6.3.2.2 Multiple feedforward

This subsection summarizes a fusion supervisor module which combines rote learning and learning from examples for the successful completion of mission M3, i.e., a collision- free navigation between the start and goal positions.

Rote learning knowledge is incorporated through a set of heuristic rules whereas the knowledge learned from examples is incorporated through a set of neural networks working in parallel. To improve the generalization capability of the neural network approach, this method uses a set of neural networks, each identifying a typical situation through pattern recognition, rather than just using one neural network as in the previous method. It is to be expected that a set of neural networks can cover a wider range of situations, and a set of heuristic rules could handle the remaining situations, thus improving operational capabilities.

While experimenting with the previous NN controller, it was observed that in most cases there were a number of special situations defined by the values of the range measurements relative to the position of the goal. The number of special situations (i.e., patterns) was determined on a trial and error fashion. Several experiments were conducted with OPMOR to group sensor measurements relative to the position of the goal. Seven groups (i.e., situations) were identified and a representative of each group was selected to train a backpropagation network. During execution, all neural networks executed in parallel with coefficients learned during the learning phase. The values of the relative contributions of each primitive behavior for each of the seven situations were determined by trial and error based on the experience with behavioral control.

The following special situations were used:

S1. The vehicle comes near to the goal position and there are no obstacles.
S2. The vehicle comes near to the goal position and there are obstacles.
S3. There are obstacles in front of the vehicle and some side of the vehicle is without obstacles.
S4. The robot is surrounded by obstacles.
S5. The robot is performing a contour following right.
S6. The robot is performing a contour following left.
S7. The goal is in sight.

Table 6.1 Relative gains of primitive behaviors for the Multiple FF supervisor

	a_1	a_2	a_3	a_4	a_5
S1	0.9	0.0	0.0	0.3	0.0
S2	0.2	0.0	0.0	0.7	0.4
S3	0.0	0.0	0.0	0.4	0.9
S4	0.0	0.0	0.0	0.1	0.9
S5	0.0	0.0	0.9	0.0	0.0
S6	0.0	0.9	0.0	0.0	0.0
S7	0.9	0.0	0.0	0.1	0.0

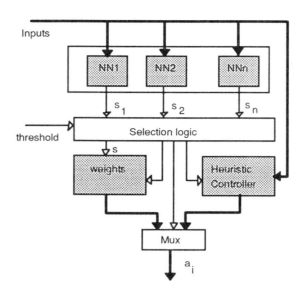

Figure 6.8 Multiple neural network architecture

As depicted in Figure 6.8, each of the set of neural networks outputs a Boolean variable indicating whether the situation given by the sensory system corresponds to one of the predefined patterns. If the situation corresponds to one of the seven predefined patterns, then the output of the controller (i.e., the action) is given by Table 6.1, otherwise the heuristic controller is used.

Simulation results show that the performance of the backpropagation neural network supervisor is not adequate, as the robot bumps into some obstacles in some situations. This problem is attributed to the training problems found with the previous neural network. The new supervisor is achieved by using the heuristic and the multiple neural network working simultaneously. The fusion supervisor includes a switching function to commute from the heuristic rules to multiple NNs according to some performance measures. Figure 6.9 illustrates the performance of this supervisor operating in a simulated environment.

This supervisor has some training advantages when it is compared with the feedforward network solution, because in the multiple feedforward network supervisor the neural networks are not trained with the outputs of the control window as in the previous supervisor. Thus, part of the errors and delays originated by the human trainer are eliminated. The main disadvantage of this supervisor derives from the fact that we are limiting the number of possible combinations of primitive behaviors to a fixed number of posibilities (seven patterns in this case) or a small number of control rules when the heuristic controller is activated. On the contrary, the outputs of the previous supervisor are extended in a continuous range of values. Another advantage of this supervisor is its robustness.

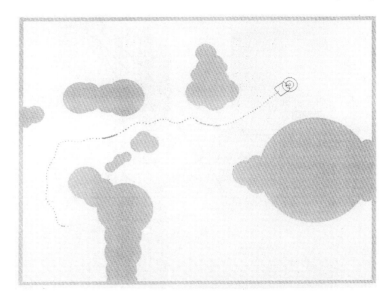

Figure 6.9 Performance of the multiple NN supervisor

6.3.2.3 Reinforcement

This subsection summarizes a fusion supervisor module which learns by exploration of the successful completion of mission M4 (surveillance of the environment), and the mission M3 (navigation between two points).

The algorithm used belongs to the method of learning by exploration or observation and is called adaptive heuristic critic (AHC). The architecture of the AHC network requires an $r(t)$ function [28], which represents the reinforcement signal to be used in the learning algorithm. We describe a reinforcement method for the task of learning how to avoid collisions while maneuvering through the environment executing missions M4 (surveillance) and M3 (go to goal).

The fusion supervisor implemented has the following inputs and outputs:

- The outputs correspond to the relative contributions of the five primitive behaviors used for the M3 mission (i.e., a_1 through a_5) or the four primitive behaviors used for the M4 mission (i.e., a_1 through a_4).
- The inputs are five ultrasound sensor measurements, where each of the measurements is the minimum distance provided by the US sensors covering certain areas. In the case of the M3 mission, another additional input to the neural network is the angle between the heading of the vehicle and the line formed by the vehicle origin and the goal position.

The neural network architecture implementing the reinforcement learning method is illustrated in Figure 6.10 and consists of:

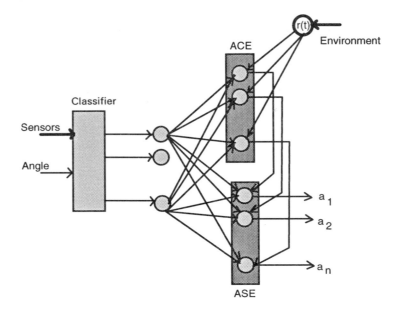

Figure 6.10 Reinforcement supervisor architecture

- An adaptive heuristic critic (AHC) network. This network uses continuous inputs and generates binary outputs, which limits the way of operating to a combination of primitive behaviors.
- An input classifier. Initially, this stage was not considered and the sensor inputs were connected directly to the AHC network. The reason for introducing the classifier was the inability of the algorithm to converge. This solution has been adopted by other researchers.

The AHC network has three layers with an input layer of sixty-four nodes, an associative critical element of five nodes, and an associative search element of five nodes. The input nodes correspond to the Boolean codification of the inputs, and the five output nodes correspond to the outputs of the supervisor.

The external reinforcement $r(t)$ provides a graded performance value describing how well the system performed. The election of a reinforcement function depends on the mission to be executed by the supervisor. We have implemented two reinforcement signals:

1. Mission M4 (surveillance).
    Because the purpose of the supervisor is to avoid collision with obstacles only, the reinforcement signal can easily be defined as: -1 in case of collision danger (we have considered collision danger when the nearest obstacle is located at less than 30 centimeters of the mobile robot), and 0 in any other case.
2. Mission M3 (go to a goal position).

In this mission, the reinforcement signal has been defined in the following way: -1 in case of collision danger or the case where the robot is not going toward the goal position when there is a free path to the goal position, and 0 in any other case.

Roughly the system works as follows. At the beginning, the robot executes random movements with $r(t)$ being evaluated continuously and used for changing the internal gains. When collision danger situations occur, the robot is returned to the position ten steps before the collision took place, and the process is continued from there. The learning process involves the association of actions with the external reinforcement (i.e., whether collision danger situations are present) for each situation. As a result of the learning process, the system learns how to navigate without collision danger situations. The number of collisions decreases as the robot goes through more learning cycles. After a sufficient number of collisions, the robot navigates with virtually no collisions.

The advantage of this unsupervised method is that there is no need for examples as it generates knowledge autonomously.

Figure 6.11 illustrates the initial collisions of the robot in an initial simulation experiment, whereas Figure 6.12 shows the path of the robot in a subsequent simulation experiment where the weights of the NN were stored at the end of each experiment and used in the following experiment. It can be seen that the number of collision danger situations decreases as the robot goes through more learning cycles. After a sufficient number of collision danger situations, the robot navigates with virtually no collision danger situations in the simulator.

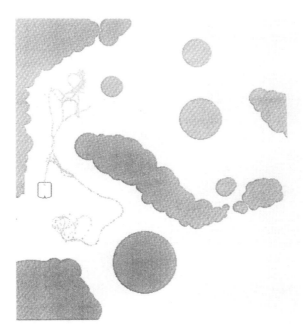

Figure 6.11 Initial collisions of a surveillance robot

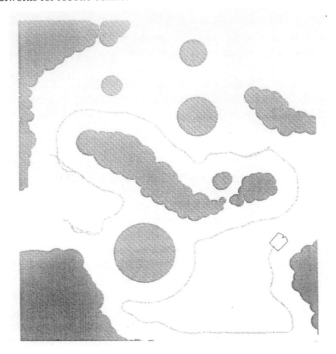

Figure 6.12 Final path of the surveillance robot

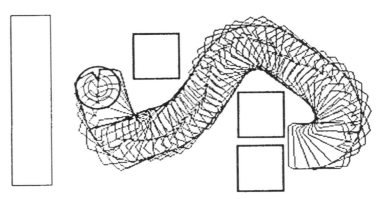

Figure 6.13 Execution of feedforward NN in the actual robot

6.4 EXPERIMENTAL RESULTS IN A REAL MOBILE ROBOT

The final step is to execute the control software on the actual robot. As noted a number of times, perhaps the main advantage of the development environment is that the control software developed under the simulator is virtually the same as that used to control the

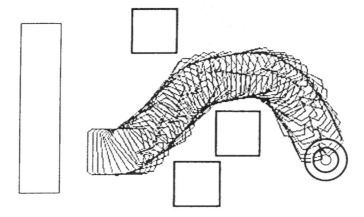

Figure 6.14 Execution of multiple NN in the actual robot

actual vehicle. In most cases, the only changes made to the control software in the simulator before it controls the actual robot are of two categories: lowering the speed of the robot, and adjusting   some parameters (e.g., the distance away from contours when following it). Other changes involve additional sensor processing (e.g., a filter) to deal with highly noisy sensors, sensor failures, multiple reflexions, and no-reflexions. The filter used with experiments with the actual robot works as follows: if a range measurement is below 30 cm (the lower limitation of the ultrasonic sensor) it is discarded and replaced by the mean of two valid measurements at each side of the sensor in question. This filter has proven practical in solving many problems dealing with bad range measurements.

The physical laboratory environment used corresponds to a small room of 6.5 m × 4.5 m where we usually put boxes as objects. The size of the room poses a number of practical problems as it does not leave enough space to maneuver. Thus the room is an excellent test bed for maneuvering in tight spots. By contrast, we did not have similar maneuvering problems in the simulation environment. As noted, by reducing the velocity of the robot and adjusting some parameters of some primitive behaviors, we were able to run all our simulation control algorithms with the real vehicle in the small room.

Figures 6.13, 6.14, and 6.15 illustrate the movement of the Robuter in the room where we have placed four boxes at different locations and orientations. There are only two differences in the control software running with the robot as compared to that running with the simulator: the actual robot speed is lower and the number of back steps after a collision danger situation is reduced from thirty to ten in the case of the reinforcement fusion supervisor.

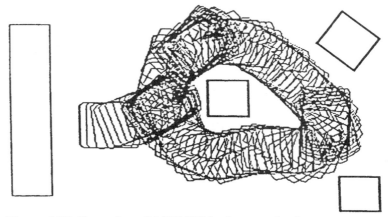

Figure 6.15 Execution of AHC NN in the actual robot

6.5 CONCLUSIONS

The previous supervisors have been developed in a simulation environment, and finally implemented on the actual robot. As noted a number of times, perhaps the main advantage of the development environment is that the control software developed under the simulator is virtually the same as that used to control the actual vehicle. In most cases, the only changes made to the control software in the simulator before it controls the actual robot are of two categories: lowering the speed of the robot, and adjusting some parameters (e.g., the distance away from contours when following it). Other changes involved additional sensor processing (e.g., a filter) to deal with highly noisy sensors, sensor failures, multiple reflexions, and no-reflexions.

Different methods have been developed to supervise the way of fusing the different behaviors. Learning algorithms involve rules, examples, advice, and solutions to similar problems. Several learning approaches were considered and three learning algorithms were implemented with a Robuter platform: learning from examples, integrating rote learning and learning from examples, and learning from explorations.

The research emphasis on autonomous learning was placed on the sensing level, which is the lowest level as it is the closest to the sensors. The primary tool used at the sensing level is the artificial neural network, with all its advantages and limitations. The NN learning paradigm takes a relatively narrow viewpoint as it primarily considers only sensor information and actions. More work needs to be done involving situations, experiences, and lessons (i.e., at higher layers). Learning paradigms that can deal with situations, experiences, lessons, mission-oriented performance levels and their integration with lower-level learning paradigms are needed, such as the neural network.

Research in learning for autonomous mobile robots must be bottom-up, starting with sensations and situations at a lower level, and incorporating experiences and lessons at higher levels. Indeed, it appears that a hierarchical learning architecture is an appropriate one, such as the one suggested by Shibata *et al.* [29]. One reason why the traditional symbolic AI research is not appropriate for autonomous learning is that it is primarily based top-down and not based on an actual sensory system.

Integrating learning and control in autonomous mobile robots is important because learning how to perform appropriate control actions rather than trying to generate these control functions to work under a wide range of circumstances seems to be an effective method. Using traditional control methodologies is difficult due to the complex nature of the robot operating environment. For most problems of practical interest the size of the working space is huge. Thus, direct trial and error methods are prohibitively expensive. Liu [14] has classified dynamical systems in terms of parameter space as $O(n)$, $O(n^2)$, etc. with a Hopfield neural network being a $O(n^2)$ system and other neural network systems being of order $O(n^3)$, where n is the dimension of the input space (e.g., the number of ultrasonic sensors in a mobile robot).

ACKNOWLEDGEMENTS

This research was   funded by the Comisión Interministerial de Ciencia y Tecnología CICYT (Projects ROB90-159, ROB91-64) and the EEC Esprit-2483 Panorama Project.

REFERENCES

[1]   T. L. Anderson and M. Donath (1990), Animal behavior as a paradigm for developing robot autonomy, in *Designing Autonomous Agents*, P. Maes (Ed.), MIT Press, pp. 145-168.

[2]   R. C. Arkin (1989), Motor schema-based mobile robot navigation, *Int. J. of Robotics Research*, **8**, 4, pp. 92-112.

[3]   R. C. Arkin (1990), Integrating, behavioral, perceptual, and world knowledge, *Reactive Navigation, Robotics and Autonomous Systems,* **6**, pp. 105-122.

[4]   R. A. Brooks (1986), A robust layered control system for a mobile robot, *IEEE J. Robotics and Automation*, **RA-2**, pp. 14-24.

[5]   R. M. H. Cheng, J. W. Xiao and S. Le Quoc (1992), Neuromorphic controller for AGV steering, *IEEE Int. Conf. on Robotics and Automation*, France, pp. 2057-62.

[6]   R. J. Clark, R. C. Arkin and A. Ram (1992), Learning momentum: On-line performance enhancement for reactive systems, *Proc. IEEE Int. Conf. on Robotics and Automation,* pp. 111-116, Nice, France.

[7]   D. Gachet, M. A. Salichs, J. R. Pimentel, L. Moreno and A. de la Escalera (1992), A software architecture for behavioral control strategies of autonomous systems, *Proc. IEEE Int. Conf. on Industrial Electronics, Control and Instrumentation (IECON'92)*, San Diego CA, pp. 1002-1007.

[8]   D. Gachet, J. R. Pimentel, L. Moreno, M. A. Salichs and V. Fernandez (1993), Neural network control approaches for behavioral control of autonomous systems, *1st IFAC Int. Workshop on Intelligent Autonomous Vehicles*, UK.

[9]   D. Gachet, J. R. Pimentel, E. A. Puente, M. A. Salichs and R. Valverde (1993), Learning behavioral control by reinforcement for an autonomous mobile robot, *Proc. IEEE Int. Conf. on Industrial Electronics, Control and Instrumentation (IECON-93)*, Hawaii, **1**, pp. 1436-41.

[10] L. P. Kaelbling and S. J. Rochestein (1990), Action and planning in embedded agents, *Robotics and Autonomous Systems,* **6**, pp. 35-48.

[11] I. S. Kweon, Y. Kuno, M. Watanabe and K. Onoguchi (1992), Behavior-based intelligent robot in dynamic indoor environments, *Proc. Int. Conf. on Intelligent Robots and Systems (IROS'92)*, Raleigh, NC, pp. 1339-1346.

[12] D. T. Lawton, R. C. Arkin and J. M. Cameron (1990), Qualitative spatial understanding and reactive control for autonomous robots, *Proc. IEEE Int. Workshop on Intelligent Robots and Systems, IROS'90,* pp. 709-714.

[13] M. A. Lewis and G. A. Bekey (1992), The behavioral self-organization of nanorobots using local rules, *Proc. Int. Conf. on Intelligent Robots and Systems (IROS'92), Raleigh, NC*, pp. 1333-1338.

[14] Y. Liu (1992), Two pattern learning algorithms using dynamical systems, *Proc. Int. Conf. on Intelligent Robots and Systems (IROS'92)*, Raleigh, NC, pp. 1363-1363.

[15] P. Maes (1990), Situated agents can have goals, *Robotics and Autonomous Systems,* **6**, pp. 49-70.

[16] P. Maes and R. Brooks (1991), Learning to coordinate behaviors, autonomous mobile robots: control, planning and architecture, IEEE Comp. Soc. Press, pp. 224-30.

[17] L. Moreno, E. Moraleda, M. Salichs and A. de la Escalera (1993), Fuzzy supervision of behavioral primitives of autonomous systems, *Proc. IEEE Int. Conf. on Industrial Electronics, Control and Instrumentation (IECON'93),* Hawaii, USA.

[18] D. W. Payton, J. K. Rosenblatt and S. M. Keirsey (1990), Plan guided reaction, *IEEE Trans. on Syst., Man., and Cybernetics*, **20**, 6, pp. 1370-1382.

[19] D. W. Payton (1990), Internalized plans: A representation for action resources, *Journal of Robotics and Autonomous Systems,* **6,** pp. 89-103.

[20] J. R. Pimentel, E. A. Puente, D. Gachet and J. M. Pelaez (1992), OPMOR: Optimization of motion control algorithms for mobile robots, *Proc. IEEE Int. Conf. on Industrial Electronics, Control and Instrumentation,* San Diego, CA, pp. 853-61.

[21] J. R. Pimentel, D. Gachet, L. Moreno and M. A. Salichs (1993), Learning to coordinate behaviors for real-time path planning of autonomous systems, *IEEE Int. Conf. on Syst., Man. and Cybernetics*, Le Touquet, France.

[22] J. R. Pimentel, D. Gachet, L. Moreno and M. A. Salichs (1993), On-line performance enhancement of a behavioral neural network controller, *Int. Workshop on Artificial Neural Networks (IWANN'93)*, Sigtes, Spain.

[23] N. S. V. Rao, E. M. Oblow, C. W. Glover and G. E. Liepins (1992), N-learners problem: Fusion of concepts, *Proc. Int. Conf. on Intelligent Robots and Systems (IROS'92)*, Raleigh, NC, pp. 1372-1379.

[24] Robosoft (1992), *Robuter User's Manual,* Robosoft, Paris, France.

[25] K. Saga, T. Sugasaka, M. Sekiguchi, S. Nagata and K. Asakawa (1992), Mobile robot control by neural networks using self-supervised learning, *IEEE Trans. on Industrial Electronics*, **39**, 6, pp. 537-542.

[26] M. A. Salichs, G. Gachet, E. A. Puente and J. R. Pimentel (1993), Reinforcement learning for behavioral navigation of mobile robots, *Proc. IEEE Int. Conf. on Industrial Electronics, Control and Instrumentation (IECON'93)*, Hawaii, USA.

[27] M. Salganicoff and R. K. Bajcsy (1992), Robotic sensory motor learning in continuous domains, *IEEE Int. Conf. on Robotics and Automation*, Nice, France, pp. 2045-2050.

[28] P. K. Simpson (1989), *Artificial neural systems: foundation, paradigms, applications and implementations*, Pergamon Press.

[29] T. Shibata, T. Fukuda, K. Kosuge, F. Arai, M. Tokita and T. Mitsuoaka (1992), Hybrid symbolic and neuromorphic control for hierarchical intelligent control, *IEEE Int. Conf. on Robotics and Automation*, Nice, France, pp. 2081-2086.

[30] P. P. van der Smagt, B. J. A. Krose and F. C. A. Groen (1992), A self-learning controller for monocular grasping, *Proc. Int. Conf. on Intelligent Robots and Systems (IROS'92)*, Raleigh, NC, pp. 177-181.

7

A neural network controller for the navigation and obstacle avoidance of a mobile robot

R. Biewald
St Petersburg State University of Electrical Engineering, Department of Automation and Control Engineering, St Petersburg, Russia [1]

7.1 INTRODUCTION

This chapter is concerned with the evaluation of a neural network (NN) based real-time navigation system (NavS) for realistic, non-holonomic vehicle models (objects with kinematic constraints, i.e. they cannot turn-in-place), assuming that only constraint global world knowledge is available.

Modern flexible manufacturing requires fast and safe operating autonomous transport systems. The performance of existing autonomous mobile robots (MRs) in speed, safety and complexity is rather poor if compared with human abilities to steer a vehicle. Applying NNs can help to emulate many aspects of human behaviour in real time. The aim of this chapter is to demonstrate how NNs may be applied in this way.

7.1.1 Mobile robot navigation
Traditional robot navigation approaches, such as potential fields [1] and graph search methods [2, 3], are based on a detailed, accurate metric description of the environment. They are difficult to implement in real time - especially if they take the kinematics constraints of the vehicle into account [4, 5]. On the other hand, fast obstacle avoidance algorithms employ idealised robot models (turn-in-place vehicles) [6]. Since the conventional approaches represent the path (or at least the robot and target position) in terms of artificial coordinate systems, they are highly vulnerable to spatial inaccuracy in sensory devices and movement actuators [7].

Nguyen and Widrow [8, 9] proposed a neural controller that steers a trailer truck

---

[1] The author is currently with Information- and NC-System GmbH, Gehringstr. 39, 13088 Berlin, Germany.

from almost any initial position to a loading dock using the truck position state to generate the next steering signal. The advantage of this approach is that it controls realistic vehicle models (highly non-linear plants with kinematic constraints), but does not take into account the current environment. Plumer [10] introduced a hybrid network architecture to permit truck control in the presence of obstacles, implementing the conventional potential field method in a parallel processing system. Although this approach solves real-time path planning for non-holonomic MRs,[2] some other drawbacks of traditional approaches still remain. Employing a more qualitative (human-like) world model seems to offer some promise of overcoming the fragility of purely spatial methods.

7.1.2 Human navigation behaviour

Humans do not employ an accurate spatial environment model - at least not in terms of a coordinate system - to fulfil the navigation task. They use only some abstract, symbolic world knowledge (associative links between places and paths [11]) to plan a route. This symbolic route description defines the necessary actions to be performed at selected, classifiable places in order to pilot a vehicle from a starting position to a destination. In addition, the steering and low-level navigation behaviour of humans, such as obstacle avoidance, is based on experience and skills rather than on high-level planning procedures (as used in common artificial intelligence robot NavS). Human travelling performance can be seen as a small set of basic visual guidance activities (such as turning right or following a road) which are adapted to current, specific environmental configurations [12]

NNs have been shown to be successful in imitating human (skilled) behaviour [13, 14]. Consequently, in our approach we suggest the implementation of these basic guidance activities into a set of NN controllers, based on the powerful control learning architecture of Nguyen and Widrow [8]. In this connection, we use as the controller input sensory information rather than the position state of the vehicle. This enables the NN controller to combine the local navigation strategy (abstract action) with obstacle avoidance.

7.2 THE NAVIGATION SYSTEM

7.2.1 Overall structure of the system

Figure 7.1 shows the structure of the overall NavS which fulfils the symbolic world description paradigm and covers all four major navigation activities: path control, obstacle avoidance (local navigation), localisation, and path planning (global navigation).

Global navigation is performed by a task and path planner that selects the most appropriate route (e.g. shortest, safest, etc., or weighted combinations of optimisation criteria) to complete the required task. A symbolic route is defined as a sequence of places

---

[2]However, the potential field method guarantees only a safe motion for a circular shape around a specified point of the vehicle, but does not explicitly take into account the kinematic constraints. Consequently, in unfavourable circumstances other parts of the vehicle may collide with obstacles.

and paths, and actions. An action describes how to get from one place to another. The decision maker, responsible for localisation, compares the route description with the current environment by a classification system using sensory information. As a result, an action corresponding to the route is chosen. Local navigation is performed by a control unit that generates the required steering signals to execute the abstract action in the specific environmental configuration using sensory data. The steering commands cause a movement of the vehicle according to its kinematics and dynamics. In this way, by transforming the steering signal into corresponding low-level motor commands, the path control is realised. At the next position the MR's sensors repeat the scanning of the environment, producing the required input information for the localisation unit and local navigator.

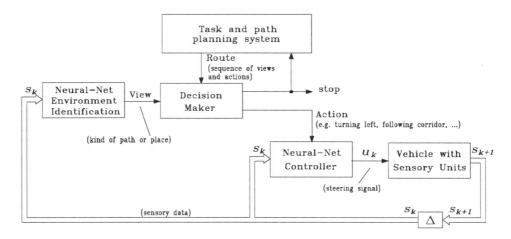

Figure 7.1 Block diagram of the navigation system

7.2.2 Symbolic world description

Most man-made environments can be classified as simple semi-regular environments as shown in Figure 7.2. Figure 7.3 represents the corresponding symbolic map, a network of paths and places [11]. The nodes represent the places and the connections are the paths. This environmental model has the advantage that it can describe relatively complex (but structured) environments with a comparatively small number of various places and paths.

The minimum set of places to describe such environments includes cross-roads and T-junctions as places, and corridors and left and right walls of various curvature as paths. The corresponding set of actions must cover 'turn left', 'turn right', and 'go ahead' at cross-roads (see Figure7.4) as well as 'follow corridor', 'follow left wall', and 'follow right wall'. In a further developmental stage of this NavS the set may be extended. Various properties of the places or paths (metrical parameters, dangerous or safe, etc.). can be stored as attributes of the nodes or connections.

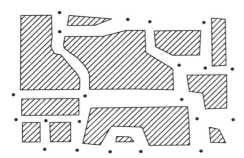

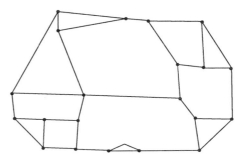

Figure 7.2 Typical man-made environment (places are selected)

Figure 7.3 The corresponding network of places and paths

Navigation may also involve environmental exploration and map building. Since the symbolic world model contains only high-level abstract spatial information, we assume that such a simple model can be built even by a human designer with little effort. Nevertheless, automated exploration can be performed using the place and path classification system. An excellent discussion of such an exploration system that forms symbolic world models is given in [15]. (One may also see the navigation approach proposed here as enabling Kuipers and Byun's qualitative topological mapping schema to cope with non-holonomic robot models).

7.2.3 Place and path classification

We used in our work ultrasonic sensors, since they are fast, cheap and for many navigation tasks satisfactorily accurate. Because, usually, up to a few tens of ultrasonic sensors are employed on MRs, it also appears reasonable to use one NN input for each physical sensor. This simplifies the system architecture, avoids pre-processing, and combines fast environmental sensing with fast parallel processing. However, this assumption does not put any restrictions on the general system architecture - any other kind of sensory data can be fed into the NNs, but it might be necessary to find other input representations or to do some pre-processing.

A simple multi-layered perceptron trained by backpropagation (BP) is used for the place and path classification. The vector of distance values to the nearest obstacles obtained by the sensors is fed directly into the NN. The binary network output vector represents the set of classifiable places and paths.

7.2.4 Local navigation and obstacle avoidance

For local navigation and obstacle avoidance an extended version of Nguyen and Widrow's motion control architecture is used. The position state of the vehicle, used as the input of the neural controller, is replaced by the sensor signals. Consequently, the controller now learns to produce steering signals while avoiding nearby obstacles. The plant represents the vehicle kinematics (and dynamics) as well as the sensory units.

The local navigator combines two kinds of input information: the sensory signals and the action selected by the decision maker. This enables the controller to perform various navigation tasks at the same environmental sample (Figure 7.4). In our approach we trained separate neural controllers for each basic action rather than defining the action as an additional network input. This reduced significantly the complexity of the task(s) to be learned by one network and achieved a higher quality of control performance. Pomerlau [13] also pointed out that it is more efficient to train several domain specific networks than to train a single network for all situations.

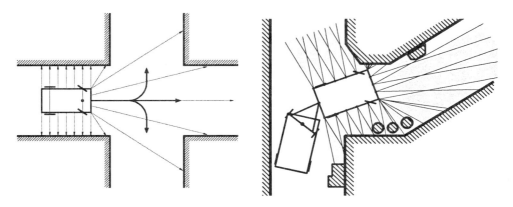

Figure 7.4 Possible motion directions on a cross-roads   **Figure 7.5 A mobile transporter in a factory environment**

In a hardware implementation the selecting function of the decision maker can be realised by loading various weight matrices into the parallel processing controller. However, it is crucial to define the right switching times since, when one controller takes over from another, instability of the system may occur. Hence, the decision maker has to make sure that at the switching moment the vehicle position is covered by the operational field of both controllers.

7.3 LOCAL NAVIGATION USING BACKPROPAGATION THROUGH TIME

The control aim of the local NavS is to generate a sequence of steering signals that applies abstract actions (extracted from a global navigation aim) to specific environmental configurations. For instance, the mobile transporter, shown in Figure 7.5, is only supplied with two items of information: a) the outputs of the ultrasonic sensors and b) it has to turn right at the coming T-junction. Thus, based on the current position relative to the nearby obstacles, it generates the appropriate steering signals that will not cause a collision in the near future. In this way, the neural controller somehow defines a (sub)optimal trajectory for the action in a specific, but classifiable, environment.

Although various controllers will be trained, the general control system structure and training procedure are the same for all of them. Training of neural controllers to perform various tasks is achieved by setting up corresponding desired responses.

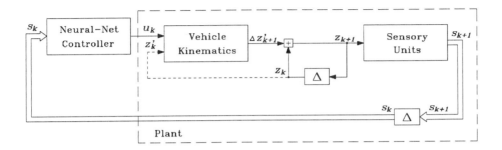

Figure 7.6 Plant details and neural controller

7.3.1 Control system structure

Figure7.6 shows the extended control system containing three two-layered feedforward nets - the NN controller, and the emulator nets for the vehicle kinematics and sensory units. For training the NN controller by supervised learning the error signal

$$\varepsilon_{u_k} = u_{k_{desired}} - u_{k_{net}} \tag{7.1}$$

for all times k is needed. Since the sequence of steering signals $u_{k_{desired}}$ that makes the vehicle follow the desired trajectory is not known, the error signals of the desired trajectory $z_{k_{desired}}$

$$\varepsilon_{z_k}^{traj} = z_{k_{desired}} - z_{k_{net}} \tag{7.2}$$

or of the desired distance values $s_{k_{desired}}$

$$\varepsilon_{s_k}^{traj} = s_{k_{desired}} - s_{k_{net}} \tag{7.3}$$

are utilised. In order to generate the error signals at the controller output, backpropagation through time (BPTT) [16] is applied. In contrast to Nguyen and Widrow, the plant was separated into two subsystems to simplify plant modelling. Before training the NN controller, two NNs have to be trained to behave like the plants using standard BP (plant identification).

The network architecture is similar to that described by Nguyen and Widrow. All nets contain an input buffer and two layers of net units - one hidden and one output layer.

The sigmoid activation function $f(x) = \tanh(x)$ is used.

7.3.2 The vehicle emulator net

The position and orientation of the employed vehicle model is uniquely characterised by a state vector of three components $z_k = \{x_k, y_k, \theta_k\}$, where x_k, y_k are the centre coordinates of the front axle and θ_k is the orientation (angle) of the vehicle (Figure 7.7). The steering signal u_k is proportional to the angle of the front wheels. Note that the position state z_k of the vehicle is only an internal representation of the plant model. Once the neural controller has learned the required task, the system operates without any artificial coordinate system.

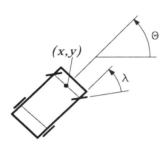

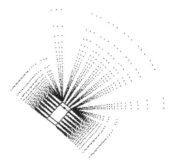

Figure 7.7 Vehicle model
($\lambda = \lambda_{max} u$)

Figure 7.8 Full range of the sensory units

In contrast to Nguyen and Widrow, the emulator network (containing twenty hidden units) is trained to predict the change of the state Δz_{k+1}, rather than the new state z_{k+1}, in order to achieve a higher prediction accuracy [17]. Furthermore, a specific representation of Δz_k is employed: $\Delta z'_k = \{p_k, o_k, \Delta\theta_k\}$, where p_k is the forward movement in the direction of θ_{k-1}, o_k is the sideward replacement, and $\Delta\theta_k$ is the orientation change. This allows training of the emulator net without feeding the current state into the network (Figure 7.9). Training is achieved by presenting randomly chosen steering signals u_k to the network while the net learns to predict $\Delta z'_{k+1}$. Note that training such an emulator is not absolutely essential, since the sensitivity matrix for the error BP during controller training could be used.

7.3.3 The environmental emulator net

The emulator net for the sensory units is trained to predict the output values s_k of each sensor i for the position states z_k in the operational range of the MR (Figure 7.10). The network emulates the environment scanned by sensors rather than the sensory units only. Therefore, for various sample environments, different emulators have to be trained. The

NN somehow learns an internal representation of the surrounding obstacles and produces a modified output for each sensor.

The vehicle is equipped with $u=25$ ultrasonic sensors (Figure 7.8). We followed an approach similar to Opitz's [18] by using a logarithmic resolution of the sensory signals to enable accurate steering control near to the obstacles. Our environment emulator nets contained 20-40 hidden units in one layer - further layers and units did not improve the training result. Training of both emulator nets required approximately 12000-25000 trials. If we achieved a root-mean-squared (rms) error of approximately 1% per output for the vehicle emulator, then for the environmental emulation results of only 4% to 10%, depending on the complexity of the environmental sample, were reached. To decrease the rms error we kept the size of the environmental samples as small as possible and restricted the possible vehicle orientations according to the action to be trained in this sample.

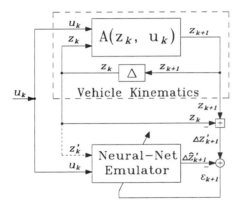

Figure 7.9 Training of the vehicle emulator

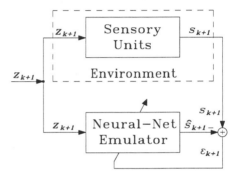

Figure 7.10 Training a neural net to store an environment

Despite the not very impressive training results, it is hoped that the environment approximation achieved is sufficiently accurate for the error BP during controller training, since for the forward pass the real environment data will be used. Once the emulator networks are trained, they can be used for controller training.

7.3.4 Controller training using position states
The objective function for training by position states is to follow a desired trajectory:

$$\mathbf{J} = E\left(\sum_{k=1}^{K}\left(c_d\left(0-d_k\right)^2 + c_\theta\left(\theta_{traj}-\theta_k\right)^2\right)\right) \tag{7.4}$$

where d_k is the distance to the desired trajectory and θ_{traj} the tangential angle of the trajectory. The constants c_d and c_θ are chosen by the designer to weigh the importance of each error component.

The desired trajectory may not be feasible, since the motion of the vehicle is restricted by kinematic constraints. An appealing property of the BPTT is that it will find a trajectory which matches the desired one as closely as possible, according to the objective function **J**. Although the designer has to specify the desired trajectories that the vehicle may follow, the optimisation properties of the BPTT provide the potential for approximate solutions.

Figure 7.11 illustrates controller training in forward (recall) phase and in backward (learning) phase. The vehicle is positioned at a random state z_k that belongs to the set of possible initial positions. The sensors then scan the environment and deliver the required inputs s_k for the controller. The controller generates the appropriate steering signal u_k, which moves the vehicle to a new position z_{k+1}, and so on. The vehicle stops either when at least one sensor detects a collision, or when the maximum number of time steps K determined by the designer has been reached.

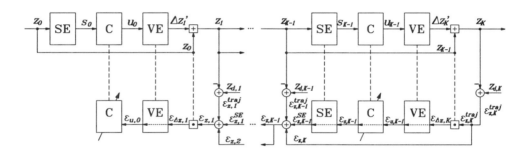

Figure 7.11 Neural net controller training using position states
(C - controller, VE - vehicle emulator, SE - sensory emulator)

In the reverse sequence, the errors are backpropagated through all nets, while the weights are adjusted only in the controller net. The portion of error generated by BP through the nets $\left(\varepsilon_{z_k}^{SE} + \varepsilon_{z_{k+1}}\right)$ can be seen as responsible for adjusting the steering commands that caused a deviation from the desired trajectory in the "near future" (trajectory optimisation). On the other hand, the error portion $\varepsilon_{z_k}^{traj}$ punishes the controller for immediate errors in tracking the reference trajectory. Note that during the forward pass real plants are used and the emulator nets are only updated in parallel in order to reduce error accumulation.

7.3.5 Controller training using distance values
One aim of the proposed NavS architecture is to reduce the designer's effort in determining the local trajectories. Learning by position state is used only for fast initial training.

Training by distance values adapts and optimises motion trajectories to specific environmental configurations and is more appropriate for generalisation to unknown environments.

The designer determines a desired output value s_i (distance) for each sensor i that should be retained during task execution. In our work we defined an interval covering the smallest and the largest distance (respectively) rather than a single distance. Thus, (7.3) becomes:

$$\varepsilon_{s_{i_k}}^{traj} = \begin{cases} \left(s_{i_{min}} - s_{i_k}\right), & \text{for} \quad s_{i_k} < s_{i_{min}} \\ \quad\quad 0, & \text{for} \quad s_{i_{min}} \le s_{i_k} \le s_{i_{max}} \\ \left(s_{i_{max}} - s_{i_k}\right), & \text{for} \quad s_{i_k} > s_{i_{max}} \end{cases} \tag{7.5}$$

In the example shown in Figure 7.12, the desired distance from the wall is only defined for the right side sensors (for the left side it is set to a value larger than the sensory scanning radius).

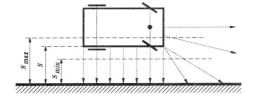

Figure 7.12 Defining the sensory error   **Figure 7.13 Approaching and following a trajectory**

Controller training using distance values is similar to the process shown in Figure 7.11, while replacing the trajectory errors of z_k by those of s_k. The objective function becomes:

$$\mathbf{J} = E\left(\sum_{k=1}^{K} \sum_{i=1}^{n} c_i \left(\varepsilon_{s_{i_k}}^{traj}\right)^2\right) \tag{7.6}$$

7.4 NEURAL NET CONTROLLERS FOR LOCAL NAVIGATION

7.4.1 Learning to keep distance

Controller training is divided into several lessons. Firstly, the controller is trained to follow a trajectory parallel to a wall (keeping the vehicle at a safe distance, Figure 7.13). On the basis of this common initial controller, several other neural controllers are trained to perform various tasks.

Since initially the steering signals of the controller are random, the motion sequence is stopped after $K=10$ time steps. The controller weights converge after approximately 100 training cycles. The controller net only contains ten hidden units. Relatively small learning rates (0.001-0.0003) are employed to avoid unstable performance.

7.4.2 Learning to follow a wall

Training by sensor distance values was employed on the next lesson. The vehicle (re)learned to follow the wall at various distances (Figure 7.14). The weight matrix of the already trained controller was used for weight initialisation and, consequently, retraining was achieved very quickly after 20-30 training cycles. For retraining to follow the wall at small distances (1/10 of vehicle length), very small learning rates ($\approx$0.0001) have to be employed, otherwise the network shows a tendency to produce oscillating steering signals (alternately hard right and left).

In order for the controller to turn the "steering wheel" to the left in time while approaching the wall, we introduced a second desired distance interval for the right front sensors. Good simulation results were achieved if the value was set to larger values than the ones needed from the geometrical viewpoint ($\approx$ 1 vehicle length). After further training, this is generalised to allow the finding of feasible motion trajectories at L-shaped wall configurations (Figure 7.15).

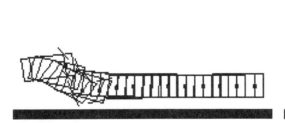

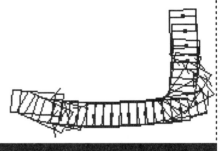

Figure 7.14 Approaching and following a trajectory at a small distance from the wall (1/5 to 1/3 of vehicle width)

Figure 7.15 Generalised wall following performance

7.4.3 Learning to turn and go straight ahead on cross-roads

Finally, different controllers were designed to perform the three standard tasks at cross-roads. At first, the NN was trained to steer the car around the right corner (Figure 7.16). The weight matrix was initialised by using the weights of the "wall following" controller. For initial training, the "moving through obstacle" approach was applied, i.e. the vehicle did not stop when colliding with an obstacle. This is a kind of unrealistic training (at least for the training of systems implemented in hardware). On the other hand, it seems useful to give the neural controller some initial knowledge through simulations, before doing the fine-tuning on real MRs.

The learning speed of this task is very impressive. As few as five training cycles were needed to achieve good "turning" performance (Figures 7.17 and 7.18). The same

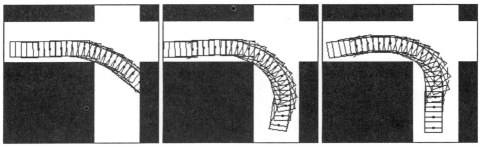

Figure 7.16	**Figure 7.17**	**Figure 7.18**
Performance	**Performance**	**Performance**
before training	**after one training cycle**	**after five training cycles**

training result can be achieved if the execution is stopped when the vehicle hits an obstacle, however over 100 training cycles were needed to achieve this performance.

An NN may be trained to perform the "turning left" task by employing the same algorithm. However, it might be simpler to use the trained "turning right" controller, and generate the appropriate steering signals by spatially mirroring the sensory inputs of the NN controller as well as changing the sign of the steering signal.

Figure 7.19 illustrates the performance of a controller that has learned to steer a car straight ahead. This NN was trained to follow a trajectory defined by the position states. By thorough observation of Figure 7.19, one can see that the car has some orientation problems at the centre of the cross-roads, since the side sensors output an "infinite" distance to the nearest obstacles. At the crossing centre, the neural controller obtains its only environmental information from the few front sensors (compare with Figure 7.4). Most probably, a more dense placement of front sensors will improve the "go ahead" performance.

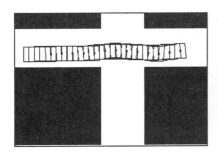

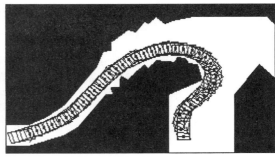

Figure 7.19 Another neural net controller that has learned to go straight ahead

Figure 7.20 Performance in unknown environment

7.4.4 Generalisation to unknown environments

Due to the generalisation capabilities of the NNs, the controllers also perform well in environments which they have never seen before. No training took place in the environment shown in Figure 7.20. The performance of the vehicle was especially impressive, since the controller employed was only trained on straight line walls. The car was able to follow curved walls and perform quite a smooth trajectory on highly irregular wall configurations.

7.5 IMPROVING TRAINING FOR LOCAL NAVIGATION

7.5.1 Environmental prediction

NN controllers can learn to perform well in a wide range of environmental configurations, even when they were only trained in a constrained set of environmental samples. However, training of emulator nets that have to learn to remember various environmental samples is not satisfactory in terms of training time and rms error. To overcome this problem one might treat the environment in the same way as vehicle kinematics: predicting the next environment view s_{k+1} (or, more exactly, the environmental variation Δs_{k+1}) is based on the current view s_k, rather than *storing* the environment.

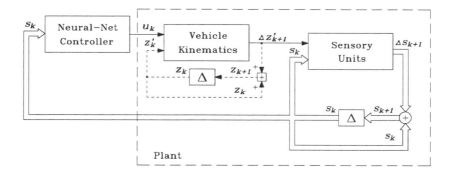

Figure 7.21 Plant details and controller for environmental prediction system

Employing the modified state change vector $\Delta z'_k = \{p_k, o_k, \Delta\theta_k\}$ has the advantages that it defines the vehicle motion uniquely (in contrast to $\Delta z_k = \{\Delta x_k, \Delta y_k, \Delta\theta_k\}$ according to the environmental prediction. The corresponding control system architecture, shown in Figure 7.21, illustrates that, for the vehicle model employed, the absolute state vector z_k may be completely excluded from the plant structure emulated by NNs.

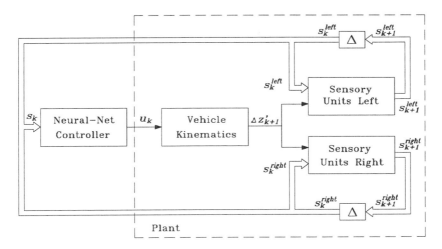

Figure 7.22 Plant details and controller for separate environment prediction (simplified)

Note that the environmental predictor net may be employed for controller training in environments in which training of the emulator has not taken place. Of course, the predictor net has to be trained initially with sufficient valid environmental data.

7.5.2 Separate environment prediction

Furthermore, we might even separate the environmental prediction for the vehicle sides (Figure 7.22), since the sensory information of one side is not needed to predict the view of the opposite side. In this manner, the complexity of the learning task is reduced and confusion from the sensory data of the opposite side is excluded. This will give some increase in prediction quality. In addition, learning to estimate future views of the environment can be constrained to one vehicle side. By spatially reflecting the net inputs and outputs on the vehicle centre line the same network can be employed to predict the environment on the opposite vehicle side.

7.6 PLACE AND PATH CLASSIFICATION

7.6.1 The view identification network

A system that is able to generalise over a wide range of noisy input vectors containing scanned distance values is needed for place and path identification and classification. NNs have been shown to work well on a large number of complex classification tasks. To enable an NN to do a good job in classifying sensory data it is important to select a suitable input representation of information.

The simplest and most preferable way of representing the sensory data to the network is to feed the vector of distance values (or proportional values) obtained by the ultrasonic sensors directly into the NN input buffer. For convenience the same vector of sensor signals s_k used for the controller network in the local navigation task is employed as input for the classification network.

The multi-layered perceptron trained by backpropagation is a classical net architecture which is mostly used for classification purposes [e.g. 19]. Since standard multi-layered feedforward NNs are also used for the NavS (motion control), the author prefers to use this kind of network in order to keep the system architecture (in hardware and software) simple.

The number M of outputs r_{m_k} of the network is equal to the number of different views to be classified. The training samples are characterised by vectors containing a one at the element of the trained view and zeros at the other elements:

$$r_{m_{k_{desired}}} = \begin{cases} 1, \ for \ m_{correct} \\ 0, \ for \ m = 1 \dots M, m \neq m_{correct} \end{cases} \tag{7.7}$$

The training of the identification NN is shown in Figure 7.23. The objective function for the training is

$$\mathbf{J} = E\left(\sum_{m=1}^{M} c_r \left(r_{m_{k_{desired}}} - r_{m_{k_{net}}} \right)^2 \right) \tag{7.8}$$

Accordingly, a typical error signal is employed for training the neural net

$$\varepsilon_{r_k} = r_{k_{desired}} - r_{k_{net}} \tag{7.9}$$

In the recall phase the net output r_{m_k} with the highest output level is simply assumed to be the right classification.

7.6.2 Desired classification areas

An important question is the definition of the best identification distance, which means the distance of the vehicle to the place at the moment of the first correct identification of the view. Of course, an identification as "early" as possible is the most preferable result. However, the probability of correct classification decreases with increasing distance to the place, simply due to the loss of analysable sensory information.

Figures 7.24 and 7.25 demonstrate that the most usable sensory information can be obtained when the top of the vehicle meets the entry of the place to be classified. On the other hand, an identification before approaching the place is a condition for good turning performance (compare Figure 7.18). Accordingly, as a good compromise we have chosen a distance of two-thirds of the vehicle length from the entry of the place as the training distance for the first correct classification (see Figure 7.26). (For high speed applications

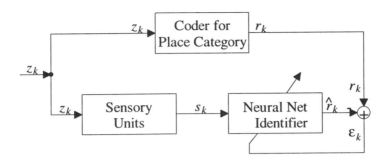

Figure 7.23 Training the place classification network

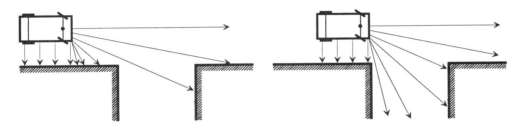

**Figure 7.24 Constrained sensory
information about the place at a far
distance from the location**

**Figure 7.25 More useful sensory
information about the place at a close
distance from the location**

requiring to reduce the vehicle speed before turning, a larger identification distance will be
needed.)

7.6.3 Training results
In our approach we trained the NN to identify seven common views that corresponded to
the highlighted training areas in the sample environment of Figure 7.26:

(a)   CO   corridor (path, way, road),
(b)   CR   X-shaped cross-roads,
(c)   TF   T-shaped cross-roads, frontal,
(d)   TL   T-shaped cross-roads, left junction,
(e)   TR   T-shaped cross-roads, right junction,
(f)   LC   left corner,
(g)   RC   right corner.

The vehicle position inside the training areas is generated randomly using a
uniform distribution. After about 20 000 training cycles in the sample environment we
achieved an average error of the classification task over all sample views of about 3.2%

while decreasing the learning rate from 0.05 to 0.003 during training. The best result was achieved for the corridor (CO) with only about 2% wrong identifications, and the poorest result was obtained for the T-shaped cross-roads, frontal (TF), having an error of about 5.5%. Wrong identifications for the last case (TF) were often the left or right corners (LC, RC), depending on the vehicle position in the training area.

The employed identification NN contains one hidden layer with thirty hidden units. Adding a second hidden layer or more hidden units did not improve the training results. Most probably, a more dense placement of the front sensors of the vehicle could improve the performance of the classification task.

Eventually, an environmental identification for the front, left and right sides by separate networks similar to those suggested in Section 7.5.2 did not improve performance in the training environment, but should improve the robustness of identification in the working environments. In this case the network only looks for the existence of a junction on the appropriate vehicle side and the resulting view is obtained by combining the output information of these three networks.

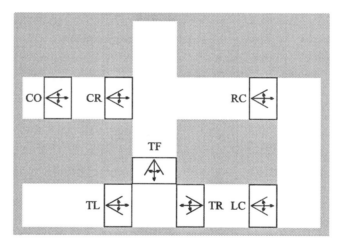

Figure 7.26 Sample environment for training the identification network with highlighted desired classification areas

7.7 A CASE STUDY

In this section we would like to demonstrate the performance of the complete navigation system in a factory environment using a relatively simple system architecture. For our study we applied only standard neural net controllers and view classifiers, trained in the general environmental samples as described in the preceding sections. There was no training of additional NNs and no extra training of the existing NNs in the chosen test environment.

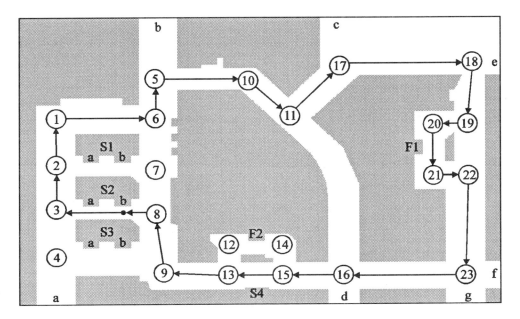

Figure 7.27 Factory environment with marked places and a sample route

7.7.1 Symbolic environmental description

The test factory environment used for the case study is shown in Figure 7.27. A network of nodes, representing the places (cross-roads, corners), and connections, representing the paths (ways, corridors), is applied for the symbolic environment description. The numbers marking the places in Figure 7.27 correspond to the rows in Table 7.1 illustrating the network description of the environment in a way that is useful for route planning algorithms. Although a correct environmental identification requires seven different views, here the environment could be described by just three kinds of places (depending on the number of junctions):

(a)   CR   X-shaped cross-roads,
(b)   T   T-shaped cross-roads,
(c)   C   corner (L-shaped cross-roads).

Column four of Table 7.1 illustrates that you will again find seven views when approaching these places from various directions.

In addition, any number of properties of nodes (places) and connections (paths) can be stored by means of the network description of the environment. This may be of assistance in checking the correctness of the navigation task carried out. We limited our extra information by storing the angular relations between the junctions of the places as well as the change of the (angular) orientation of a path between two places. The approximate length of a path, the (maximum, minimum and average) width of a path (corridor), and other specifications might be applied as other properties supporting an exact vehicle navigation.

Completing such a symbolic description of a factory environment is simple, since it does not need exact measurement and can be done by any technician in a few hours. Small and medium obstacles (pallets with workpieces, other vehicles) can be neglected. In comparison, producing a relatively precise environmental map required for the most exact path planning algorithms [3, 4, 6, 10] needs a lot more effort and storing capacity.

7.7.2 Route planning

Having defined the graph (network) of nodes and connections, symbolically representing the environment, the shortest graph (path) can be searched by some standard search technique, such as A^* search [20] or its modifications. When using these search methods for graphs, representing environmental models with high resolution, such as occupancy grids [3,4,7], the search algorithms may become very complex and the calculation time may reach unacceptable values (see also [17]). Since the symbolic world representation keeps the graphs simple, the calculation time and complexity remains small.[3]

In our case study the chosen task for the MR is to bring a pallet with a workpiece from store S2b to the fabrication centre F1, after that it should return to the store S2 while passing store S4. The planned route is marked in Figure 7.27 by the arrows. Table 7.2 contains the corresponding route information for the decision maker. The turning angle is calculated as

$$\vartheta_{turn} = 180° - \sum_{i=n_{coming}}^{n_{leaving}-1} \vartheta_{junc_i} \qquad (7.10)$$

7.7.3 Decision making

The task of the decision maker is to define the actual position of the MR in relation to the desired route and to choose the appropriate controller for the required local navigation task by comparing the symbolic route description with the identified places. In a simple form the decision maker may work by an algorithm given in Table 7.3.

To improve the robustness of this algorithm we have implemented a few additional rules that should minimise the number of wrong decisions. There are two main problems.

Firstly, the identification network only supplies a correct result while working in the standard environment in about 96% of the cases. Applying this network in a non-standard environment, the performance of the place classification will probably deteriorate. Therefore we only accept a place as identified if the same result from the identification network is obtained twice in a row. This rule will exclude most of the false classifications.[4]

---

[3] For our system we did not implement a graph search algorithm, since the route planning for such a typical factory environment can be done very easily by hand.

[4] In the standard environment the probability of correct identifications is now higher at 99.8%.

Table 7.1 Symbolic description of the factory environment in Figure 7.27

No. of place	Kind of place	No. of connect. place	View from connect. place	Angle to next junction ϑ_{junc}	Orientation change of path ϑ_{path}
1	C	6	LC	270°	0°
		2	RC	90°	0°
2	T	7	TF	90°	0°
		1	TL	180°	0°
		3	TR	90°	0°
3	T	8	TF	90°	0°
		2	TL	180°	0°
		4	TR	90°	0°
4	T	9	TF	90°	0°
		3	TL	180°	0°
		a	TR	90°	0°
5	T	10	TF	90°	0°
		b	TL	180°	0°
		6	TR	90°	0°
6	T	5	TR	90°	0°
		1	TF	90°	0°
		7	TL	180°	0°
7	T	6	TR	90°	0°
		2	TF	90°	0°
		8	TL	180°	0°
8	T	7	TR	90°	0°
		3	TF	90°	0°
		8	TL	180°	0°
9	T	13	TR	90°	0°
		8	TF	90°	0°
		4	TL	180°	0°
10	C	5	RC	135°	0°
		11	LC	225°	0°
11	T	17	TF	90°	0°
		10	TL	180°	0°
		16	TR	90°	- 45°
12	C	14	LC	270°	0°
		13	RC	90°	0°
13	T	15	TR	90°	0°
		12	TF	90°	0°
		9	TL	180°	0°
14	C	12	RC	90°	0°
		15	LC	270°	0°
15	T	16	TR	90°	0°
		14	TF	90°	0°
		13	TL	180°	0°
16	CR	23	CR	90°	0°
		11	CR	90°	45°
		15	CR	90°	0°
		d	CR	90°	0°
17	T	18	TF	90°	0°
		c	TF	135°	0°
		11	CR	135°	0°
18	T	e	TL	180°	0°
		17	TR	90°	0°
		19	TF	90°	0°
19	T	18	TR	90°	0°
		20	TF	90°	0°
		22	TL	180°	0°
20	C	19	LC	270°	0°
		21	RC	90°	0°
21	C	22	RC	90°	0°
		20	LC	270°	0°
22	T	19	TR	90°	0°
		21	TF	90°	0°
		23	TL	180°	0°
23	CR	f	CR	90°	0°
		22	CR	90°	0°
		16	CR	90°	0°
		g	CR	90°	0°

Furthermore, in some situations it might be very difficult for the identification network to distinguish two similar places. For example, a right corner may look like a T-shaped cross-roads (frontal) if an additional obstacle is situated at the corner entry on the left side (compare Figure 7.27, place 1). However, this "wrong" identification has no influence on the correct execution of the navigation task, since the vehicle will turn right in both cases. Accordingly, we have defined a list of compatible places that are also accepted as correct when actually looking for another place. As Table 7.4 indicates, recognising a compatible place also depends on the local navigation task required at this place.

Table 7.2 Symbolic description of the route shown in Figure 7.27

No.	No. of place	View of place	Turning angle ϑ_{turn}	Reqr. ac-tion	Con-trol-ler used	Orient. change of path ϑ_{path}
1	S2b			(S)	R	0°
2	3	TF	-90°	R	R	0°
3	2	TR	0°	S	L	0°
4	1	RC	-90°	R	R	0°
5	6	TF	90°	L	L	0°
6	5	TR	-90°	R	R	0°
7	10	RC	-45°	R	R	0°
8	11	TL	90°	L	L	0°
9	17	CR	-90°	R	R	0°
10	18	TR	-90°	R	R	0°
11	19	TR	-90°	R	R	0°
12	20	LC	90°	L	L	0°
13	F1					
14	21	LC	90°	L	L	0°
15	22	TF	-90°	R	R	0°
16	23	CR	-90°	R	R	0°
17	16	CR	0°	S	S	0°
18	15	TR	0°	S	L	0°
19	S4					
20	13	TR	0°	S	L	0°
21	9	TR	-90°	R	R	0°
22	8	TL	90°	L	L	0°
23	S2b					

Table 7.3 A simple algorithm for decision making

```
start
    choose controller for local navigation
    while not end of route
        get next place from route plan
        go until next route place is identified
        choose controller for local navigation
    endwhile
stop
```

Secondly, the performance of the identification network is not known when the vehicle is actually passing the place, since any training did not take place there. Consequently we designed a rule that forces the decision maker to ignore the output of the identification network for a defined period of time after a correct place identification. In our approach this period of time is specified for turning actions as long as the vehicle does not change its orientation more than two-thirds of the required turning angle given in the route plan for this place. For straightforward motions this time corresponds to a travel distance which on average is needed for passing places by such a kind of motion[5].

The corresponding modified algorithm of the decision maker is given in Table 7.5.

---

[5] In our example environment this distance was choosen as eleven movement steps (= 5.5 m), since the vehicle travels 0.5 m each time step. The average width of a corridor is 4 m , the vehicle length is designed as 3 m and the second correct identification can be expected at 1.5 m distance from the place.

Table 7.4 List of compatible places

		Place	\multicolumn{6}{c}{Identified Place [accepted as a compatible place if one of the enumerated navigation actions is required at the place searched (R - turning right, S - going straight ahead, L - turning left)]}					
		Place	CR	TR	TF	TL	RC	LC
Searched		CR	R/S/L	S/L	R/L	R/S	R	L
place		TR	R/S	R/S	R	-	R	-
		TF	R/L	R	R/L	L	R	L
Real		TL	S/L	-	L	S/L	-	L
place		RC	R	R	R	-	R	-
		LC	L	-	L	L		L

7.7.4 Local navigation

Just a small set of two different NN controllers was employed in the factory environment:

(a) a controller performing the task of turning to the right on cross-roads, and

(b) a controller performing the task of going straight ahead on cross-roads.

Since the turning controller also demonstrates a suitable performance while following a corridor on the right side (compare Figure 7.20) we applied it as a general local navigation controller.

For all tasks requiring motions on the left side, including turning to the left, the required performance was obtained using the same general (turning to the right) controller by spatially reflecting the sensory inputs as well as the steering signal. Accordingly, the decision maker has to select between three basic local navigation tasks:

(a)   S   going straight ahead,

(b)   R   going to the right,

(c)   L   going to the left.

7.7.5 Simulation results

Figure 7.28 demonstrates the navigation of an MR through the factory environment based on the route plan given in Table 7.2. The environment contains some additional obstacles, such as other parked vehicles and stored pallets and workpieces. Table 7.6 (see appendix) represents a detailed description of the task performed, including the route plan, the outputs of the identification network r_k and the decision maker d_k as well as the orientation of the vehicle θ_{veh_k} at every step k. This gives the opportunity for detailed performance analysis of the navigation system introduced.

Table 7.5 An extended algorithm for decision making

---

start
    choose controller for local navigation
    while not end of route
        if change of path orientation $\vartheta_{path_l} \neq 0°$ then
            go until corridor (CO) is twice identified
            save θ_{veh_k} as $\theta_{veh_{save}}$
            go until $\frac{2}{3}\text{sign}(\vartheta_{path_l})(\theta_{veh_k} - \theta_{veh_{save}}) > \left| \vartheta_{path_l} \right|$
        endif
        get next place and path from route plan
        go until next route place r_{route_l} is twice identified
        choose controller for local navigation
        save θ_{veh_k} as $\theta_{veh_{save}}$
        go until $\frac{2}{3}\text{sign}(\vartheta_{turn_l})(\theta_{veh_k} - \theta_{veh_{save}}) > \left| \vartheta_{turn_l} \right|$
    endwhile
stop

Generally, the navigation task was fulfilled successfully; the vehicle followed the route accurately, passed all the necessary places and returned to the starting position while avoiding (nearly all) additional unexpected obstacles. Although the identification network did not always output the correct place information at all vehicle positions (Table 7.6) the extended algorithm enabled the decision maker to overcome these faults and to choose the necessary navigation controllers at the corresponding places. Even the difficult job at the fabrication centre F1 requiring a very short switching interval between the different places was mastered quite well. However, in two situations the navigation algorithm proposed did not perform the local navigation task satisfactorily.

Firstly, at place number 1 the MR collides with another parked vehicle. This is caused by the method of training of the turning controller. The controller has learned its task at standard environments (standard shaped cross-roads) only and does not expect additional obstacles after turning which noticeably narrow the remaining width of the pathway. Further training of this controller in non-standard environments should correct this poor performance.

The second problem is connected with passing and entering very small corridors as required at stores S1 and S2. Although the MR is able to travel through this small corridor when the motion is started inside the corridor, it cannot be excluded that the vehicle will collide with the wall. Furthermore, the navigation controller does not seem to be able to steer the vehicle to enter such a small pathway. Additional training of the controller might also help to enable the neural controller to fulfil this task in a more efficient way. However, it does not appear very practicable to train one NN to perform a

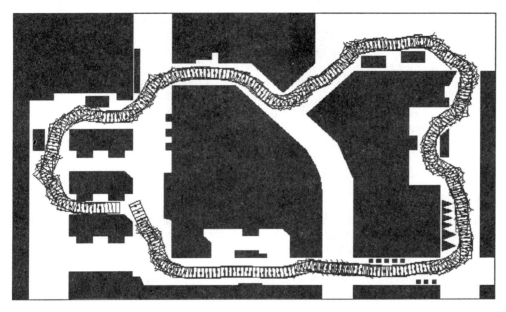

Figure 7.28 Performance of the mobile robot in the factory environment following a given route

very wide range of tasks, since the controller training is based on keeping a (predefined) safety distance from all obstacles. An alternative solution could be the design of an additional neural local navigation controller for entering and moving in small pathways. Unfortunately, this suggestion requires the ability of the neural identification to distinguish these small corridors from paths with normal width before entering them.

At some places the steering actions of the MR do not appear straightforward for an experienced car driver. For instance, the insufficient performance of the vehicle is very conspicuous when passing T-shaped cross-roads (place number 2) and after passing the X-shaped cross-roads (place number 16). These navigation actions are caused by the control algorithm chosen using only two different navigation strategies - the left- and right-orientated ones. To pass a TR-cross-roads straight ahead the vehicle motions are controlled by the distance to the left wall. If the distance to the left wall is larger than the trained one the vehicle will approach the left side, although a smooth motion straight ahead would be more convenient. Improved steering performance can again be achieved by employing specialised neural navigation controllers.

The last example also shows, that in environments with much free space (e.g., with very wide corridors) the use of specialised controllers is absolutely necessary. Since the employed controllers are only trained for narrow environments, the navigation system may become unstable while switching from the right-orientated to the left-orientated controller. For instance, doing this during the vehicle motion between place 17 and 18 in our sample environment would provoke an unpredictable steering performance.

Finally, we can establish that, as expected, the motion performance of the MR at the standard shaped places (e.g. place 11) is smooth and safe while at non-standard locations (e.g. place 1) such an ideal performance could not be achieved. Note again that no training took place in the factory environment sample. On the other hand, if an MR is designed to work in a constrained environment (with a limited number of possible routes), then no argument exists against "updating" the neural controller performance by additional training in the real environment and against the improvement of identification and local navigation in this way.

7.8 CONCLUSIONS

7.8.1 Summary

An application of NNs to steering and navigation of non-holonomic MRs is introduced here. Industrial MRs are highly non-linear plants with kinematic constraints whose control is difficult to realise with conventional methods. BPTT is an NN approach which has been successfully applied to control such complex plants. BPTT has the appealing property of training NNs by self-learning in order to generate control signals that combine immediate and long-term error correction. In this way, with very little design effort, feasible motion trajectories for plant constraints are found.

Based on sensory information, an NN controller learns to steer a vehicle around obstacles while performing abstract navigation actions extracted from a global task planner. No pre-processing of the sensory data is required. Consequently, the NN performs a complex sensor-action behaviour. Inexpensive ultrasonic sensors may be employed, but any other sensory data are also applicable.

The same sensory data as used for local navigation can be employed for the simple location classification system. The case study has shown that the developed system works well in previously unknown environments, even with relatively poor identification results due to the diversity of actual environmental views.

Due to the generalisation capabilities of NNs, much of the high-level sequential planning activities (typical for artificial intelligence-based navigation in robotics) were replaced by neural adaptive control mechanisms. Ergo, the complexity of the employed world model is reduced to a symbolic representation, since local trajectory "planning" is completed by the NN controller. The world model is robust with respect to small and medium environmental changes, since obstacle avoidance is performed automatically during local path execution. All this, together with an implementation in parallel hardware, reduces the processing time dramatically and guarantees real-time performance. In addition, the possibility of importing different weight matrices into one unique hardware allows us to perform various tasks while keeping the control system structure very simple.

However, in its present form the NavS is restricted to man-made semi-regular environments. The system may also have some orientation problems in environments with much free space. Additional wide range perception systems should help to overcome these limitations.

Nguyen and Widrow reported about 1000 to 2000 training cycles per lesson needed to train the NN controller for the truck. In the present work NN controller convergence could be achieved in at most 150 training cycles, since, firstly, the forward kinematics of the employed vehicle model are much simpler, and secondly, some kind of reference trajectory could be utilised for training (whereas for the truck backer-upper only the desired final state was known). However, it seems that the spread of input information over 25 sensory units also plays an important part in enabling fast controller training.

7.8.2 Future research

Although the author claims that the NavS introduced is applicable to realistic vehicle models, the completed computer simulation still contains some simplifications. This makes it difficult to implement the NavS of the given form in a real MR. Future work will be concerned with removing these simplifications. NN controllers that output additional control signals should be investigated: at least a signal controlling the vehicle speed as well as the steering angle.

The simulation results obtained give us enough evidence to believe that the system can easily be extended to control an MR with a trailer, as illustrated in Figure 7.5. Introducing an NN controller with memory (recurrent net) may avoid the need to equip the trailer with sensors. Finally, future work will also be concerned with further investigations of the environmental prediction nets. It is possible that such nets might be useful for the avoidance of movable obstacles.

7.9 ACKNOWLEDGEMENTS

The main part of this research was carried out at the Control System Centre, University of Manchester Institute of Science and Technology, UK. The author would like to thank Prof. N. Munro and all the other staff of the centre for permitting and arranging for him to complete this research. Special thanks are due to Prof. P.E. Wellstead for his kind advice and encouragement in the main phase of this work.

Further thanks to Prof. D.Ch. Imaev of the St Petersburg State University of Electrical Engineering for his support in preparing this paper.

APPENDIX

Table 7.6 comprises a detailed description of the navigation task shown in Figure 7.28. The columns contain information about the current movement step (No.), the place number corresponding to Figure 7.27 (Pl.), the employed local navigation controller (C), the next (expected) place and the local navigation action required at this place given by the route plan of Table 7.2 (Route), the output of the place classification network (Ident.) and the (angular) orientation of the vehicle (Orient.). The first and second correct place identifications (including compatible places) are marked by x and X respectively in column "Ident".

Table 7.6 Navigation of an MR - data for Figure 7.28

No.	Pl.	C	Route	Ident.	Orient.	No.	Pl.	C	Route	Ident.	Orient.	No.	Pl.	C	Route	Ident.	Orient.
0	S2b	R	TF-r	CO	180.00	38		R	(-60°)	CR	75.19	76	5	R	-	CO	26.41
1	S2b	R	TF-r	CO	181.25	39		R	-	CR	68.17	77	5	R	-	CO	36.04
2	S2b	R	TF-r	CO	181.23	40		R	-	RC	61.39	78	5	R	-	RC	27.93
3		R	TF-r	CO	185.04	41	1	R	-	RC	56.89	79	5	R	-	CO	22.53
4		R	TF-r	CO	184.50	42	1	R	-	RC	52.21	80	5	R	-	CO	18.61
5		R	TF-r	CO	183.44	43	1	R	-	RC	49.11	81	5	R	-	CO	17.28
6	S2a	R	TF-r	CO	178.16	44	1	R	-	RC	45.84	82	5	R	-	CO	14.76
7	S2a	R	TF-r	TR-x	174.81	45	1	R	-	RC	41.90	83	5	R	-	CO	11.25
8	S2a	R	TF-r	RC-X	172.48	46		R	-	RC	36.52	84	5	R	-	CO	9.89
9		R	(-60°)	TF	174.31	47		R	-	RC	29.66	85		R	-	CO	6.69
10		R	-	TF	172.11	48		R	TF-l	RC	20.34	86		R	-	CO	4.63
11		R	-	TF	169.70	49		R	TF-l	CO	10.43	87		R	RC-r	CO	3.04
12	3	R	-	TF	164.37	50		R	TF-l	CO	0.81	88		R	RC-r	CO	1.13
13	3	R	-	TF	158.63	51		R	TF-l	TR	-7.62	89		R	RC-r	CO	-1.55
14	3	R	-	TF	152.10	52		R	TF-l	CO	-9.66	90		R	RC-r	CO	-1.72
15	3	R	-	TF	143.77	53		R	TF-l	TR	-5.68	91		R	RC-r	CO	-0.00
16	3	R	-	RC	134.91	54		R	TF-l	TR	-3.29	92		R	RC-r	RC-x	-3.58
17		R	-	RC	125.73	55		R	TF-l	CO	-1.45	93		R	RC-r	RC-X	-8.40
18		R	-	RC	116.54	56		R	TF-l	CO	-1.59	94		R	(-30°)	RC	-9.95
19		R	TR-s	TR-x	108.26	57		R	TF-l	CR-x	-0.55	95		R	-	RC	-13.01
20		R	TR-s	TR-X	107.13	58		R	TF-l	TF-X	-4.57	96		R	-	CO	-16.08
21		L	(1)	RC	117.81	59		L	(+60°)	TF	5.21	97	10	R	-	CO	-19.65
22		L	(2)	RC	125.71	60		L	-	TF	14.52	98	10	R	-	CO	-23.24
23		L	(3)	RC	132.87	61		L	-	TF	23.69	99	10	R	-	CO	-29.02
24	2	L	(4)	RC	122.06	62	6	L	-	LC	30.95	100		R	-	CO	-33.73
25	2	L	(5)	RC	111.05	63	6	L	-	LC	36.16	101		R	-	TL	-37.58
26	2	L	(6)	RC	100.25	64	6	L	-	TL	25.34	102		R	TL-l	CO	-38.71
27	2	L	(7)	RC	89.42	65	6	L	-	LC	33.42	103		R	TL-l	TL-x	-40.28
28	2	L	(8)	LC	78.66	66	6	L	-	LC	40.48	104		R	TL-l	TL-X	-42.08
29		L	(9)	CO	67.76	67		L	-	LC	46.90	105		L	(+60°)	LC	-40.51
30		L	(10)	TL	73.59	68		L	-	LC	51.86	106		L	-	LC	-41.43
31		L	RC-r	CO	78.75	69		L	TR-r	LC	55.57	107		L	-	LC	-31.45
32		L	RC-r	CO	79.15	70		L	TR-r	CR-x	59.59	108	11	L	-	LC	-22.97
33		L	RC-r	TR-x	79.00	71		L	TR-r	CR-X	64.58	109	11	L	-	TF	-15.50
34		L	RC-r	CO	76.21	72		R	(-60°)	TR	54.61	110	11	L	-	TF	-8.89
35		L	RC-r	LC	76.38	73		R	-	TF	45.93	111	11	L	-	TF	-1.74
36		L	RC-r	CR-x	81.81	74		R	-	LC	37.98	112		L	-	LC	4.70
37		L	RC-r	CR-X	82.93	75	5	R	-	LC	30.73	113		L	-	LC	13.14

Table 7.6 cont'd

No.	Pl.	C	Route	Ident.	Orient.
114		L	CR-r	LC	21.14
115		L	CR-r	LC	28.40
116		L	CR-r	CO	34.63
117		L	CR-r	TR-x	39.97
118		L	CR-r	TR-X	42.22
119		R	(-30°)	TR	33.12
120		R	-	TL	26.66
121		R	-	LC	26.56
122		R	-	TL	29.60
123		R	-	LC	33.66
124		R	-	CR	37.49
125	17	R	-	LC	39.56
126	17	R	-	LC	41.79
127	17	R	-	TL	40.76
128	17	R	-	CR	43.07
129	17	R	-	TL	40.71
130	17	R	-	TL	40.43
131	17	R	-	CR	43.58
132	17	R	-	CR	54.79
133	17	R	-	CR	65.60
134	17	R	-	CR	56.24
135	17	R	-	CR	47.30
136	17	R	-	CR	38.77
137	17	R	-	CR	31.64
138		R	-	CR	25.56
139		R	-	CR	21.24
140		R	-	CR	19.34
141		R	-	TL	14.92
142		R	TR-r	TL	10.06
143		R	TR-r	TL	5.25
144		R	TR-r	TL	0.66
145		R	TR-r	TL	1.80
146		R	TR-r	CR-x	7.42
147		R	TR-r	TR-X	5.51
148		R	(-60°)	TR	5.39
149		R	-	TR	8.14
150		R	-	TR	11.15
151		R	-	TR	7.35
152		R	-	CR	2.48
153		R	-	CR	-1.66
154		R	-	CR	-4.80
155	18	R	-	CR	6.01
156	18	R	-	CR	-2.47
157	18	R	-	CR	-10.67
158	18	R	-	CR	-19.43
159	18	R	-	TL	-26.05
160	18	R	-	CR	-32.84
161	18	R	-	LC	-38.30
162	18	R	-	LC	-42.15
163	18	R	-	LC	-43.10
164	18	R	-	CO	-31.84
165	18	R	-	TF	-21.09
166	18	R	-	TF	-29.27
167		R	-	TF	-36.45
168		R	-	RC	-26.31

No.	Pl.	C	Route	Ident.	Orient.
169		R	-	RC	-33.67
170		R	-	RC	- 40.49
171		R	-	RC	-
50.33 172		R	TR-r	RC-x	-60.37
173		R	TR-r	RC-X	-70.47
174		R	(-60°)	RC	-80.60
175		R	-	TR	-90.71
176		R	-	CO	-98.18
177		R	-	TR	-92.34
178	19	R	-	TR	-101.12
179	19	R	-	TR	-92.92
180	19	R	-	TR	-102.36
181	19	R	-	TR	-109.46
182	19	R	-	TR	-100.07
183	19	R	-	RC	-91.85
184		R	-	TR	-102.30
185		R	-	TR	-111.25
186		R	-	TF	-118.35
187		R	-	TF	-125.35
188		R	LI-l	RE	-133.11
189		R	LI-l	LI-x	-142.79
190		R	LI-l	TL-30	-150.56
191		L	(+60°)	LC	-140.56
192		L	-	LC	-132.92
193		L	-	RC	-127.93
194	20	L	-	RC	-125.93
195	20	L	-	LC	-126.95
196	20	L	-	LC	-124.71
197	20	L	-	LC	-120.63
198	20	L	-	LC	-113.32
199		L	-	LC	-105.55
200		L	-	LC	-99.63
201		L	-	LC	-95.42
202	F1	L	LI-l	LC-x	-90.32
203	F1	L	LI-l	LC-X	-83.71
204		L	(+60°)	TF	-77.00
205		L	-	LC	-68.32
206	21	L	-	TL	-58.63
207	21	L	-	CR	-48.93
208	21	L	-	CR	-39.94
209		L	-	CR	-32.04
210		L	TF-r	TF-x	-23.30
211		L	TF-r	TF-X	-21.45
212		R	(-60°)	RC	-33.46
213		R	-	RC	-43.60
214		R	-	CO	-53.48
215	22	R	-	CO	-62.85
216	22	R	-	CO	-71.23
217	22	R	-	TR	-70.47
218	22	R	-	CO	-64.37
219	22	R	-	CO	-69.04
220	22	R	-	CO	-60.68
221	22	R	-	CO	-69.98
222		R	-	CO	-78.58
223		R	-	CO	-70.23

No.	Pl.	C	Route	Indent.	Orient.
224		R	-	CO	-78.46
225		R	CR-r	CO	-86.06
226		R	CR-r	CO	-76.93
227		R	CR-r	CO	-83.65
228		R	CR-r	CO	-90.39
229		R	CR-r	CO	-91.91
230		R	CR-r	CO	-82.59
231		R	CR-r	CO	-78.00
232		R	CR-r	CO	-76.06
233		R	CR-r	TR-x	-81.40
234		R	CR-r	CR-X	-84.27
235		R	(-60°)	CR	-84.88
236		R	-	CR	-89.34
237		R	-	CR	-87.69
238	23	R	-	CR	-97.70
239	23	R	-	CR	-107.08
240	23	R	-	CR	-116.28
241	23	R	-	TF	-123.74
242	23	R	-	TF	-130.39
243		R	-	TF	-134.00
244		R	-	LC	-139.21
245		R	CR-s	RC	-144.41
246		R	CR-s	TF	-149.87
247		R	CR-s	CO	-157.84
248		R	CR-s	CO	-164.51
249		R	CR-s	CO	-169.31
250		R	CR-s	CO	-171.59
251		R	CR-s	CO	-172.24
252		R	CR-s	CO	-172.69
253		R	CR-s	CO	-170.36
254		R	CR-s	CO	-169.08
255		R	CR-s	CO	-171.06
256		R	CR-s	CO	-171.85
257		R	CR-s	CO	-174.39
258		R	CR-s	CO	-173.73
259		R	CR-s	CO	-176.66
260		R	CR-s	TR-x	-175.32
261		R	CR-s	CO	-179.52
262		R	CR-s	TR-x	-173.25
263		R	CR-s	CO	-181.14
264		R	CR-s	TR-x	-184.75
265		R	CR-s	LC	-184.00
266		R	CR-s	CR-x	-187.33
267		R	CR-s	CR-X	-186.97
268		S	(1)	CR	-177.21
269		S	(2)	CR	-186.42
270		S	(3)	CR	-176.76
271	16	S	(4)	CR	-186.89
272	16	S	(5)	CR	-176.69
273	16	S	(6)	TL	-187.34
274	16	S	(7)	TR	-176.73
275	16	S	(8)	TL	-187.31
276		S	(9)	TL	-176.71
277		S	(10)	CO	-187.41
278		S	TR-s	CO	-176.76

Table 7.6 cont'd

No.	Pl.	C	Route	Ident.	Orient.
279		S	TR-s	CO	-187.05
280		S	TR-s	TR-x	-195.55
281		S	TR-s	CO	-184.56
282		S	TR-s	CO	-191.53
283		S	TR-s	CO	-190.93
284		S	TR-s	CO	-185.85
285		S	TR-s	TR-x	-181.93
286		S	TR-s	TR-X	-174.54
287		L	(1)	TR	-166.20
288		L	(2)	TR	-163.00
289		L	(3)	TR	-165.10
290	15	L	(4)	CO	-167.72
291	15	L	(5)	CO	-169.53
292	15	L	(6)	CO	-172.08
293	15	L	(7)	CO	-173.85
294	15	L	(8)	CO	-176.17
295		L	(9)	CO	-175.00
296		L	(10)	CO	-175.75
297		L	TR-s	CO	-177.76
298		L	TR-s	CO	-178.27
299	S4	L	TR-s	CO	-178.33
300	S4	L	TR-s	TR-x	-177.63

No.	Pl.	C	Route	Ident.	Orient.
301		L	TR-s	TR-X	-180.46
302		L	(1)	TR	-181.04
303		L	(2)	TR	-181.50
304		L	(3)	TR	-182.17
305	13	L	(4)	TL	-181.39
306	13	L	(5)	CO	-180.31
307	13	L	(6)	CO	-180.03
308	13	L	(7)	CO	-179.18
309	13	L	(8)	CO	-178.71
310		L	(9)	CO	-178.36
311		L	(10)	CO	-179.10
312		L	TR-r	CO	-180.25
313		L	TR-r	CO	-180.08
314		L	TR-r	CO	-179.63
315		L	TR-r	CO	-180.98
316		L	TR-r	TL	-181.35
317		L	TR-r	RC-x	-181.66
318		L	TR-r	TR-X	-183.49
319		R	(-60°)	TR	-193.94
320		R	-	TR	-187.31
321		R	-	TR	-197.64
322	9	R	-	TR	-207.41

No.	Pl.	C	Route	Ident.	Orient.
323	9	R	-	TR	-197.86
324	9	R	-	TR	-207.56
325	9	R	-	TR	-216.68
326	9	R	-	CR	-224.75
327		R	-	TF	-232.16
328		R	-	TF	-239.08
329		R	TL-l	TF-x	-245.03
330		R	TL-l	CO	-250.89
331		R	TL-l	CR-x	-257.04
332		R	TL-l	TL-X	-262.06
333		L	(-60°)	TL	-250.96
334		L	-	CO	-240.59
335		L	-	CR	-231.30
336		L	-	RC	-225.00
337		L	-	RC	-235.82
338		L	-	RC	-246.79
339	8	L	-	TL	-245.03
340	8	L	-	RC	-247.53
341	8	L	-	LC	-244.86
342	8	L	-	TF	-245.39
343	8	L	-	TF	-245.13

REFERENCES

[1]   Warren, C.W. (1990). "A technique for autonomous underwater vehicle route planning", *IEEE Jour. of Oceanic Engineering*, 15(3):199-204.

[2]   Lozano-Perez, T., and R.H. Taylor (1989). "Geometric issues in planning robot tasks", in M. Brady (editor), *Robotics Science* , MIT Press, Cambridge, MA.

[3]   Zhu, D. and J.-C. Latombe (1991). "New heuristic algorithms for efficient hierarchical path planning", *IEEE Trans. on Robotics and Automation*, 7(1):9-19.

[4]   Tournassoud, P. (1988). "Motion planning for a mobile robot with kinematic constraint", in J.-D. Boissonnat, J.-P. Laumond (editors), *Geometry and robotics*, Lecture Notes in Computer Science, Springer, New York, , 391, 150-171.

[5]   Shiller, Z., and Y.-R. Gwo (1991). "Dynamic motion planning of autonomous vehicles", *IEEE Trans. on Robotics and Automation*, 7(2):241-249.

[6]   Borenstein, J., and Y. Koren (1991). "The vector field histogram - fast obstacle avoidance for mobile robot", *IEEE Trans. on Robotics and Automation*, 7(3):278-288.

[7]   Elfes, A. (1989). "Using occupancy girds for mobile robot perception and navigation", *Computer (IEEE)*, 22(6):46-58.

[8]   Nguyen, D.H., and B. Widrow (1990). "Neural networks for self-learning control systems", *IEEE Control System Magazine*, 10(3):18-23 (or (1991) *Int. Jour. of Control*, 54.2(6):1439-1451).

[9]   Nguyen, D.H., and B. Widrow (1989). "The truck backer-upper: An example of self-learning in neural networks", *Proc. of IJCNN-89*, 2:357-363.

[10] Plumer, E.S. (1992). "Neural network structure for navigation using potential fields", *Proc. of IJCNN-92*, 1:327-332.

[11] Kuipers, B.J. (1982). "The 'map in the head' metaphor", *Environment and Behaviour*, 14:202-220.

[12] Gibson, J.J. (1979). *The Ecological Approach to Visual Perception*, Houghton Mifflin Company, Boston.

[13] Pomerlau, D.A. (1991). "Rapidly adapting artificial neural networks for autonomous navigation", in R.P. Lippmann, J.E. Moody, and D.S. Touretzky (editors), *Advances in Neural Information Processing Systems 3*, Morgan Kaufmann, San Mateo, CA, 429-435.

[14] Shepanski, J.F., and S.A. Macy (1988). "Teaching artificial neural system to drive: manual training techniques for autonomous systems", in D.Z. Anderson (editor), *Advances in Neural Information Processing Systems*, American Institute of Physics, New York, 693-700.

[15] Kuipers, B.J., and Y.-T. Byun (1988). "A robust, qualitative method for robot spatial reasoning", *Proc. of AAAI-88*, 774-779.

[16] Werbos, P.J. (1990). "Backpropagation through time: What it does and how to do it", *Proceedings of the IEEE*, 78(10):1550-1560.

[17] Biewald, R. (1992). "Application of neural networks for steering and navigation of non-holonomic mobile robots", MSc Thesis, UMIST, Control Systems Centre.

[18] Opitz, R. (1990). "Das Lernfahrzeug: Neural network application for autonomous mobile robots", in R. Eckmiller (editor), *Advanced Neural Computers*, Elsevier Science Publisher B.V. (North-Holland), 373-379.

[19] Gorman, R.P., and T.J. Sejnowski (1988). "Analysis of hidden units in a layered network trained to classify sonar targets", *Neural Networks*, vol. 1, pp. 75-89.

[20] Nilson, N.J. (1980). *Principles of Artificial Intelligence*, Palo Alto, CA: Tioga.

8

An ultrasonic 3-D robot vision system based on the statistical properties of artificial neural networks

Sumio Watanabe
Ricoh Information and Communication R&D Centre, 3-2-3, Shin-Yokohama, Kohoku-Ku, Yokohama, 222 Japan

8.1 INTRODUCTION

Recently, several models for artificial neural networks have been devised based on the nonlinear and parallel properties found in biological neural networks. These models have been successfully applied to systems performing such tasks as pattern recognition, robotic control, and the prediction of time sequences. In order to support and enhance such systems, it is becoming more important to clarify what artificial neural networks learn, and how efficiently they learn it. We now need the theoretical foundation of artificial neural networks to establish neural information processing as part of engineering.

As is shown in [1-3], learning in artificial neural networks is equivalent to the statistical maximum likelihood method. By using this equivalence, a statistical approach can be introduced to analyze the learning structure of artificial neural networks. For example, it is known that artificial neural networks learn the relation between inputs and outputs by approximating the probability distribution on the input/output space [2], and that a quantitative relation between the empirical error and the prediction error is derived from the statistical asymptotic theory [4].

As the basic structure of neural network learning is being clarified, the mathematical design methods for the optimal neural networks are being developed by using statistical tools. Nowadays, the heuristic know-how for better generalized learning is being replaced by the theoretically best design method for the minimum prediction error, with the result that one can automatically choose the optimal neural network model for an arbitrary given task.

In this article, we introduce two statistical design methods for artificial neural networks: the first is the automatic model selection algorithm for the minimum prediction error, and the second is the pattern identification method by which a neural network can reject unknown inputs. To illustrate the effectiveness of the proposed methods, we apply them to real problems: ultrasonic image reconstruction and ultrasonic object identification.

Ultrasonic imaging has been studied in the machine vision field because 3-D images of objects can be obtained directly even in dark or smoky environments. However, as of yet, it has seldom been used in the practical object recognition system because of its low image resolution. This is due to a combination of factors, including long wavelength, a limited number of receivers, small apertures, attenuation due to propagation through air, and the nonlinear response of an ultrasonic receiver.

In spite of such difficulties, it is known that ultrasonic imaging can be applied to a practical 3-D vision system with the aid of an artificial neural network [5][6]. A neuro-ultrasonic 3-D visual sensor is now being used in the automatic object recognition system of lens production lines [7]. In this article, we show that such a neuro-ultrasonic visual sensor is improved by using statistical design methods without any heuristic know-how, which means that artificial neural networks can become established as part of engineering.

8.2 THE EMPIRICAL ERROR AND THE PREDICTION ERROR

In this section we first introduce a basic relation between the empirical error and the prediction error, and then consider its application to artificial neural networks. The concept of information criteria is modified so that both the optimal model and the optimal parameter can be simultaneously obtained.

8.2.1 Parametric estimation and information criterion

Let us consider how a machine estimates the relation between the K-dimensional input space and the L-dimensional output space. Suppose that the training samples $\{x_t, y_t\}_{t=1}^{N}$ are independently taken from the environmental probability density $Q(x,y)$ on the direct product of the input space and the output space, where N is the number of training samples. Our purpose is to estimate both the probability density on the input space and the conditional probability density function

$$Q(x) = \int Q(x, y)dy \tag{8.1}$$

$$Q(y|x) = \frac{Q(x, y)}{Q(x)} = \frac{Q(x, y)}{\int Q(x, y)dy} \tag{8.2}$$

In this article, we call the former the occurrence probability, and the latter the inference probability.

Figure 8.1 shows the basic interaction between an environment and a learning machine. First, we consider how an environmental probability distribution is estimated by a function approximation neural network given by a parametric probability density

$$P(w,\sigma, y|x) = \frac{1}{(2\pi\sigma^2)^{L/2}} \exp\left(-\frac{\|y - \varphi(w;x)\|^2}{2\sigma^2}\right) \tag{8.3}$$

In equation (8.3), $\varphi(w;x)$ is a function from the input space to the output space with a parameter w, which is realized by a neural network, σ is the standard deviation of outputs, and $\|\cdot\|$ is the Euclidean norm on the output space. In other words, we estimate $Q(y/x)$ by assuming that an output y is the sum of a function $\varphi(w;x)$ of an input x and a random noise ε

$$y = \varphi(w;x) + \varepsilon$$

where ε is taken from the normal distribution with the average 0 and the variance σ^2 (Figure 8.2). It follows that

$$\varphi(w;x) = \int yP(w;y|x)dy$$

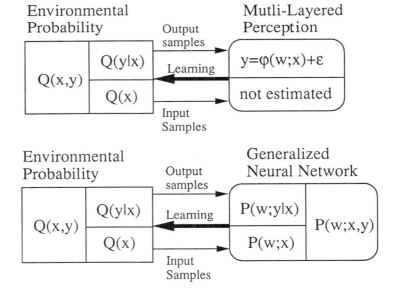

Figure 8.1 Probabilistic framework for artificial neural networks
(The first half shows how a function approximation model learns the regression function.
The second half shows how a probabilistic model estimates the environmental distribution.
The first is a special case of the second)

For a given parameter w, the empirical error is defined by

$$E_{emp}(w) = \frac{1}{N} \sum_{i=1}^{N} \| y_i - \varphi(w; x_i) \|^2 \qquad (8.4)$$

The prediction error is defined by the expectation error for the testing sample,

$$E(w) = \iint \| y - \varphi(w; x) \|^2 \, Q(x, y) dx dy \qquad (8.5)$$

We understand the optimal parameter for generalization as the parameter that minimizes the prediction error. If the number of training samples is as large as infinity, then $E_{emp}(w)$ is almost equal to $E(w)$, with the result that the optimal parameter for the minimum prediction error can be found by minimizing the empirical error. However, we usually have a finite number of training samples, and the parameter that minimizes the empirical error does not always minimize the prediction error.

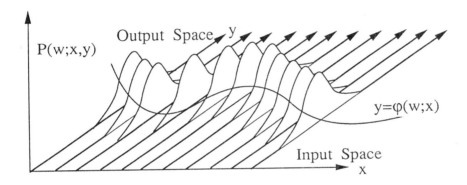

Figure 8.2 A probabilistic density realized by the multi-layered perceptron

Much research in mathematical statistics has been devoted to finding the best solution to such a problem. The basic result is as follows. When the number of training samples goes to infinity, the optimal parameter for the empirical error is asymptotically subject to the normal distribution whose variance is in inverse proportion to the number of training samples. From such a property, we can derive a quantitative relation between the empirical error and the prediction error. If the parameter w^* that minimizes the empirical error is found, then the following equations hold (proved by Akaike [5]):

$$< E_{emp}(w^*) >= \left(L - \frac{F(w^*)+1}{N} \right) \sigma_0^2 + o\left(\frac{1}{N} \right)$$

$$< E(w^*) >= \left(L + \frac{F(w^*)+1}{N} \right) \sigma_0^2 + o\left(\frac{1}{N} \right)$$

where $F(w^*)$ is the number of free parameters in w^*, σ_0^2 is the variance of the true inference probability, and $o(1/N)$ is a smaller order term which satisfies $No(1/N) \to 0$ when $N \to \infty$. Note that w^* depends on the training samples, hence w^* is a random variable. In the above equation, $< \cdot >$ means the expectation value for all pairs of training samples which are independently taken from $Q(x,y)$. Figure 8.3 shows the relations among three kinds of errors and the number of training samples. The true error $L\sigma_0^2$ is equal to the variance of the true inference probability $Q(y|x)$. From the above two equations, we have a relation between the empirical error and the prediction error:

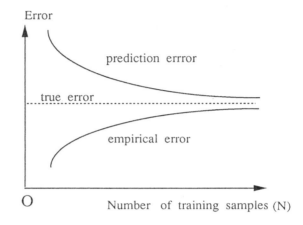

Figure 8.3 Empirical error, true error, and prediction error

$$< E(w^*) > = \left\{ 1 + \frac{2(F(w^*)+1)}{NL} \right\} < E_{emp}(w^*) > + o\left(\frac{1}{N}\right) \qquad (8.6)$$

In practical applications, only one pair of training samples is usually given; therefore, the average $< \cdot >$ cannot be calculated. However, the optimal model for the minimum prediction error can be found by minimizing the criterion

$$I(w^*) = \left\{ 1 + \frac{A(F(w^*)+1)}{NL} \right\} E_{emp}(w^*) \qquad (8.7)$$

where $I(w^*)$ is called Akaike's information criterion (AIC) [4], where $A = 2$. By using AIC, we can choose the optimal model that minimizes the prediction error as the expectation value for training samples. However, it is known that the probability of selecting the true model does not converge to one, even when the number of training samples goes to infinity. This problem is called the inconsistency of the AIC. To improve on such a problem, Schwarz's Bayes information criterion ($A = \log(N)$) [8] is proposed, which coincides with the simplest form of Rissanen's minimum description length (MDL) [9]. BIC has consistency, but it is not ensured to be optimal for the minimum prediction error.

8.2.2 Multi-layered perceptron and modified information criterion

To realize the parametric function $\varphi(w; x)$, let us consider the three-layered perceptron. Of course, another family of functions, for example a linear sum of polynomial functions, is available. However, it is known that the three-layered perceptron can approximate any continuous function more efficiently than any other linear sum of fixed basis functions if the dimension of the input space is larger than 2 [10]. The three-layered perceptron $\varphi(w; x) = \{\varphi_i(w; x)\}_{i=1}^{L}$ from the K-dimensional input space to the L-dimensional output space is defined by

$$\varphi_i(w; x) = \rho(w_{i0} + \sum_{j=1}^{H} w_{ij}\rho(w_{j0} + \sum_{k=1}^{K} w_{jk}x_k)) \tag{8.8}$$

where $w = \{w_{i0}, w_{ij}\}$ is a set of bias parameters and weight parameters, and $\rho(x)$ is a sigmoidal activation function, which is defined by $\rho(x) = 1/(1 + e^{-x})$.

The optimal parameter for the minimum empirical error is usually searched by the steepest descent method, which is called the error backpropagation, or simply BP,

$$\Delta w_{ij} = -\eta \frac{\partial E_{emp}(w)}{\partial w_{ij}} \tag{8.9}$$

where η is some constant value that determines the learning speed. It should be emphasized that BP minimizes only the empirical error. In this article, we propose the steepest descent method for the prediction error,

$$\Delta w_{ij} = -\eta \frac{\partial E(w)}{\partial w_{ij}} \approx -\eta \frac{\partial I(w)}{\partial w_{ij}} \tag{8.10}$$

Note that the optimal model, which is selected by using the information criterion, is the pruned network (Figure 8.4).

When the information criterion is applied, first the parameters which minimize the empirical errors are calculated for all possible models, and then the model that minimizes the information criterion is selected. However, in the case of neural networks, it is difficult to find the optimal parameter even for each model because of nonlinearity. The steepest descent for the information criterion can overcome such a problem, and both the optimal model and parameters can be found simultaneously. But another problem remains. The information criterion $I(w)$ is not a differentiable function for w.

In order to make the information criterion differentiable, let us introduce a modified information criterion [11], which is defined by

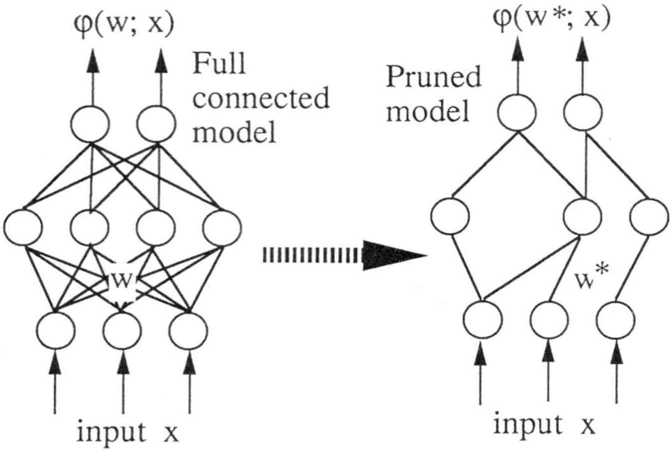

Figure 8.4 Optimization by a modified information criterion
(Both the optimal model and the optimal parameters can be chosen by using a modified
information criterion)

$$I_\alpha(w) = \left\{1 + \frac{A(F_\alpha(w) + 1)}{NL}\right\} E_{emp}(w) \qquad (8.11)$$

where $F_\alpha(w)$ is

$$F_\alpha(w) = \sum_{i=1}^{L}\sum_{j=0}^{H} f_\alpha(w_{ij}) + \sum_{j=1}^{H}\sum_{k=0}^{K} f_\alpha(w_{jk}) \qquad (8.12)$$

where $f_\alpha(x)$ is a function which satisfies the following three conditions,
    (1) The function $f_0(x)$ is 0 if $x = 0$, or 1 if otherwise.
    (2) The function $f_\alpha(x)$ is differentiable for $\alpha > 0$,
    and if $\alpha \to 0$, then $f_\alpha(x) \to f_0(x)$ pointwisely.
    (3) If $|x| \le |y|$, then $0 \le f_\alpha(x) \le f_\alpha(y) \le 1$.

For example, a function $1 - \exp(-x^2/(2\alpha^2))$ satisfies these conditions. Note that $F_0(w)$
is equal to the freedom of the parameters. Based on these conditions, we can prove the
following equation [12]:

$$\lim_{\alpha \to 0} \min_w I_\alpha(w) = \min_w I(w) \qquad (8.13)$$

This equation shows that the optimal parameter for the minimum prediction error can be
found by minimizing the modified information criterion while controlling α as $\alpha \to 0$
(Figure 8.5). The steepest descent for the modified information criterion is given by

$$\Delta w_{ij} = -\eta \frac{\partial I_\alpha (w)}{\partial w_{ij}}$$

(8.14)

$$= -\eta' \sum_{i=1}^{N} \left\{ \frac{\partial}{\partial w_{ij}} \| y_i - \varphi(w; x_i) \|^2 - \frac{A \hat{\sigma}^2}{N} \frac{\partial F_\alpha}{\partial w_{ij}} \right\}$$

where η' is a constant which is in proportion to η and

$$\hat{\sigma}^2 = \frac{1}{N} \sum_{i=1}^{N} \| y_i - \varphi(w; x_i) \|^2$$

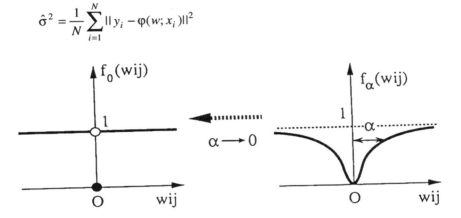

Figure 8.5 How to control α
(The parameter α plays the same role as temperature in the simulated annealing)

is the estimated variance, which is updated at the beginning of every new training cycle. By using this training method, the weights which are not necessary to approximate the target function gradually converge to zero. In other words, this algorithm can be understood as the pruning algorithm for the minimum prediction error. Note that, if the estimated variance is equal to zero, then this method is equivalent to the usual error backpropagation.

To obtain a neural network which generalizes the training samples, many heuristic methods have been proposed. The best known employs the modified error

$$E'(w) = E_{emp}(w) + \lambda C(w)$$

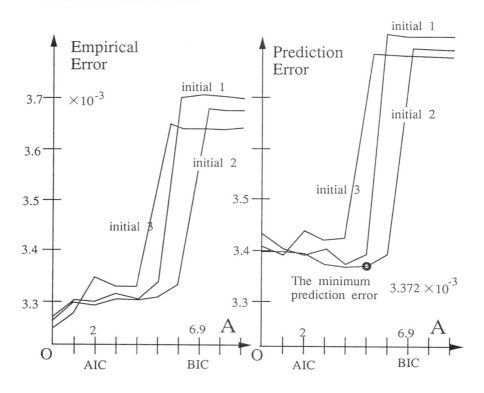

Figure 8.6 An experimental result
(The empirical errors and the prediction errors based on the modified information criterion
are shown)

where $C(w)$ is a complexity term of the neural network [13] which is usually determined by heuristic know-how, and λ is some constant determined by trial-and-error. In statistics, this method is referred to as maximum *a posteriori* estimation, but it is difficult to design *a priori* the probability density $\exp(-\lambda C(w))$ in real applications. The modified information criterion can be understood as the optimal controlling method λ and $C(w)$.

In order to illustrate how the modified information criterion affects learning, we give a simple example. One thousand input/output samples were taken from the neural network given in Figure 8.7 (1). A neural network with ten hidden units was used to estimate the averaged output of the inference probability. Figure 8.6 shows experimental results for the relation among the empirical errors, prediction errors, and several values of A. Three different lines correspond to three different initial weight values. If A is taken to be larger, then more weight parameters are pruned. If A is taken to be zero, then no parameter is pruned. Figure 8.7, (2) and (3) respectively, show the networks trained by using $A = 2$ and $A = 5$. The optimal A is conjectured to be found between AIC and BIC,

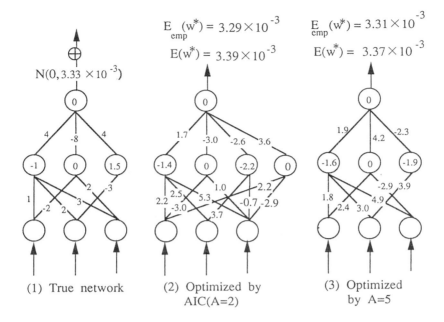

$$E_{emp}(w^*) = 3.29 \times 10^{-3}$$

$$E(w^*) = 3.39 \times 10^{-3}$$

$$E_{emp}(w^*) = 3.31 \times 10^{-3}$$

$$E(w^*) = 3.37 \times 10^{-3}$$

$$N(0, 3.33 \times 10^{-3})$$

(1) True network

(2) Optimized by AIC(A=2)

(3) Optimized by A=5

Figure 8.7 Models and parameters optimized by the modified information criterion

but, theoretically speaking, it has not yet been determined. In this example, the true model was chosen when $A = 5$.

8.3 APPLICATION TO ULTRASONIC IMAGE RECONSTRUCTION

In this section, we consider an application of the multi-layered perceptron to ultrasonic image reconstruction. Figure 8.8 shows the ultrasonic imaging system, which consists of a transmitter, an ultrasonic receiver array, and an object to be recognized. Figure 8.9 is the photograph of the system, which is now being used on the practical production line in a lens factory.

The transmitter illuminates the object with an incident wave, emitted at $t = 0$. The location of the transmitter is $\mathbf{r}_0 = (x_0, y_0, z_0)$. The sound pressure $P_{in}(\mathbf{r}, t)$ of the incident wave at location $\mathbf{r}$ and time t is then given by

$$P_{in}(\mathbf{r}, t) = \Theta(ct - |\mathbf{r} - \mathbf{r}_0|) \exp(j\mathbf{k}_{in} \cdot (\mathbf{r} - \mathbf{r}_0) - j\omega t), \tag{8.15}$$

where c is the sound velocity, ω is the angular frequency, $j = \sqrt{-1}$, $\Theta(x)$ is a Heaviside function, $\mathbf{k}_{in} = (k \sin\theta, 0, -k\cos\theta)$, k is the wave number, and θ is the illuminating angle. Let $\xi = \xi(x, y)$ and $z = \zeta(x, y)$ be the reflection coefficient and the surface function of the object. The 3-D shape of the object is defined by

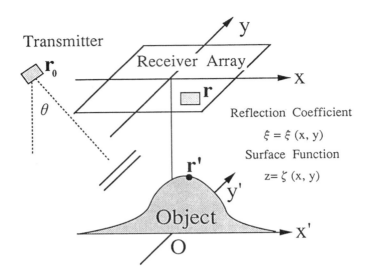

Figure 8.8 Arrangement of an ultrasonic transmitter, a receiver array, and an object to be recognized

$$f(x, y, z) = \xi(x, y)\Theta(\zeta(x, y) - z) .$$ (8.16)

Then the 3-D image can be reconstructed from the observed pressure $P(\mathbf{r},t)$ of the scattered wave at the receiver's location $\mathbf{r}$ and time t by the following equation [5].

$$f(x, y, z) = \frac{kz^2}{\pi} \left| \iint Q(\mathbf{r}', z) \exp\left(j\frac{k}{r'}(xx'+yy') \right) dx' dy' \right|$$ (8.17)

where

$$Q(\mathbf{r}', z) = \frac{P(\mathbf{r}', (r'+r_0 - 2z\cos\theta) / c) \cdot (z + z'\cos\theta)}{\{(x'-r'\sin\theta)^2 + y'^2 +(z+r'\cos\theta)^2\}r'^3 \exp(jkr')}$$

Note that the reconstructed 3-D image $f(x,y,z)$ is the Fourier transform of the function $Q(\mathbf{r}',z)$. If the area of the receiver array is taken to be larger, then a higher resolution image can be obtained. But in the real imaging system, the integration in eq.(8.17) is replaced by the finite sum of the receiver points, resulting in a distorted reconstructed image. Because the high-frequency coefficients of the image is lost, the shortest resolvable length is in inverse proportion to the width of the receiver array (Nyquist's sampling theorem), and it is an ill-posed problem to restore the complete image from finite information. Moreover, the ultrasonic receiver has a nonlinear time-delayed response, because it observes the sound pressure by using resonance. Lastly, noises are introduced into both ultrasonic waves and the system. These are the reasons why it has been difficult

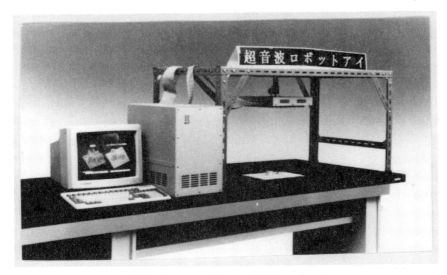

Figure 8.9 A neuro-ultrasonic 3-D visual sensor

to apply ultrasonic imaging to 3-D object recognition. It is difficult to describe the image degradation process. Artificial neural networks are expected to solve such an ill-posed, nonlinear, and probabilistic problem caused by the unknown process.

Figure 8.10 shows the three-layered perceptron used in an image reconstruction experiment. The number of input units is nine, and the input vector consists of nine values for the neighborhood with three by three pixels. The number of output units is one. In other words, this neural network was used as an adaptive nonlinear filter for image reconstruction. Three images, A, B, and C in Figure 8.11 were used as training input samples. As one image was made of 32 by 32 pixels, the number of training samples was 3072. The desired output image for the training sample was made by measuring the real size of the object.

The number of hidden units was set as fifteen at the start of the training. The initial weight and bias parameters were independently taken from the uniform probability density on the interval $[-0.1, 0.1]$. During training, the parameters were optimized by using the modified information criterion. The number of all training cycles was set as $n_0 = 30000$, and α was controlled as

$$\alpha(n) = 3.0 \frac{n_0 - n}{n_0} + 0.01$$

where n is the number of training cycles.

The first half of Figure 8.12 shows the images reconstructed by the least square method, or the usual backpropagation ($A = 0$), minimum AIC ($A = 2$), and minimum MDL ($A = \log(N)$). In the reconstructed image by AIC or MDL, rather less noise can be seen, and a tail of the character R resolved. The latter half of Figure 8.12 shows a 3-D image of the spanner in Figure 8.11 which is reconstructed by integrating 2-D images.

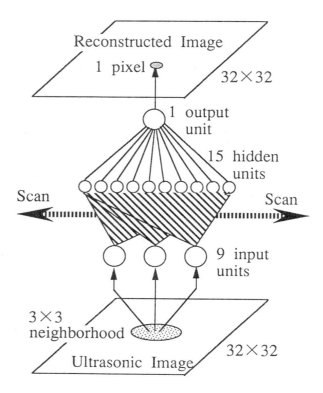

Figure 8.10 An ultrasonic image restoration method using a neural network

Some super-resolution techniques have been studied in the image restoration field. Conventional methods, however, assume that the image degradation process is completely clarified. As described above, it is difficult to clarify the nonlinear probabilistic process of ultrasonic image degradation in the real problem. An artificial neural network can estimate such a nonlinear process, and realizes the optimal nonlinear inverse filter. The modified information criterion enables us to design an optimal nonlinear filter automatically.

8.4 A PROBABILITY COMPETITION NEURAL NETWORK

In the above discussion, we showed that the output of the multi-layered perceptron (MLP) is equal to the regression function, namely, the estimated average value of the inference probability. In the pattern classification problem, the regression function is equal to the Bayesian *a posteriori* probability of a given input [14]. That is to say, if the MLP sufficiently approximates the regression function, then it realizes the optimal classification rule with the smallest error probability.

Figure 8.11 Object samples
A, B, and C were used as training samples, and other objects were testing samples

However, as is often reported, the MLP cannot identify a given input, and cannot say whether a given input is known or unknown. The MLP can classify the input vectors into categories, but it cannot say whether a given input is contained in learned concepts or not.

In practical applications, unknown samples are often inputted. For example, a character recognition system should reject unknown codes, a voice recognition system must extract only human speech from noisy sounds. In this section, we consider a new model which can overcome such problems.

Let X and Y be an input space and an output space, respectively, and $Q(x,y)$ be a probability density function from which training samples are taken. Remember that the multi-layered perceptron estimates only the expectation value of the inference probability $Q(y|x)$, but it does not estimate the occurrence probability $Q(x)$ (Figure 8.1, first half). However, in order to estimate how often a given input x occurs during training, the occurrence probability should be estimated. In this section, let us consider a novel neural network model by which both the occurrence probability and the inference probability can be estimated. Since $Q(x,y) = Q(x)Q(y|x)$, we estimate the simultaneous probability density $Q(x,y)$ by using a parametric probability density function (Figure 8.1, latter half).

Let $P(w;x,y)$ be a probability density function on the direct product $X \times Y$ of the input space and the output space, where w is a parameter to be optimized. In this section, a probability competition neural network [2] is proposed to estimate the simultaneous probability density. Let us consider the following probability density function on the direct product of the input space and the output space:

$$P(w;x,y) = \frac{1}{Z(\theta)} \sum_{h=1}^{H} \frac{1}{(2\pi\rho_h^2)^{K/2}(2\pi\sigma_h^2)^{L/2}} \exp\left(-\frac{\|x-x_h\|^2}{2\rho_h^2} - \frac{\|y-y_h\|^2}{2\sigma_h^2} + \theta_h\right)$$

where H is the number of hidden units, $w = \{x_h, y_h, \rho_h, \sigma_h, \theta_h; h = 1,2,\ldots,H\}$ is a set of parameters to be optimized by training, and

$$Z(\theta) = \sum_{h=1}^{H} \exp(\theta_h)$$

Namely, the above probability density function is a convex combination of normal distributions (Figure 8.13), whose averages and deviations are optimized by training. If the above parameters can be sufficiently optimized so that $P(w;x,y) \approx Q(x,y)$, then we have both the estimated occurrence probability $P(w;x)$ and the estimated inference probability $P(w;y|x)$ by

$$P(w;x) = \int P(w;x,y)dy$$

$$P(w;y|x) = \frac{P(w;x,y)}{P(w;x)}$$

(8.18)

The occurrence probability $P(w;x)$ estimates how often a given input occurs, or how familiar the neural network is with a given input. The average of the inference probability $E(w;x) \equiv \int yP(w;y|x)dy$ estimates the output expected for a given input, or *a posteriori* probability of the categories for a given input. It should be emphasized that, if the estimated occurrence probability is very small, then the inference is ill-defined because

$$E(w;x) = \frac{\int yP(w;x,y)dy}{\int P(w;x,y)dy} \approx \frac{0}{0}$$

which means that an artificial neural network may output a nonsense value for an unknown input. In such a case, a given input should be rejected as an unknown input. If the estimated occurrence probability is large, then the inference probability is well-defined and its output can be expected to be precisely one. In other words, the estimated occurrence

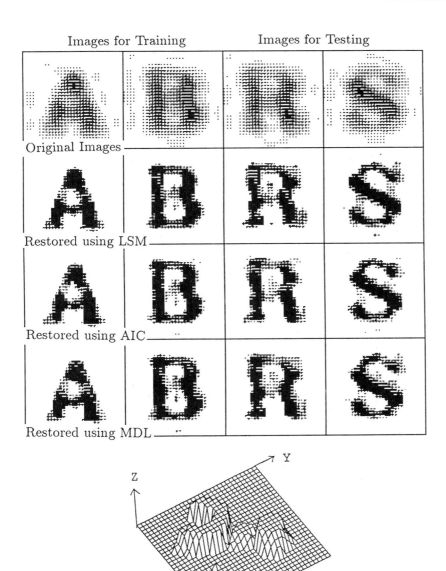

Figure 8.12 Reconstructed images by a multi-layered perceptron
(Ultrasonic images are restored by a multi-layered perceptron which is trained by a
modified information criterion. A 3-D image can be reconstructed by integrating 2-D
images)

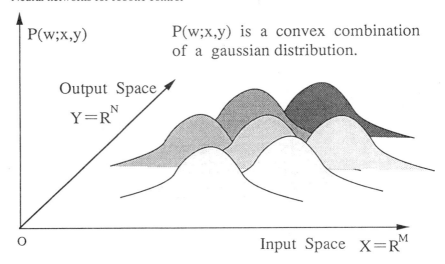

$P(w;x,y)$ is a convex combination of a gaussian distribution.

Output Space $Y=R^N$

O   Input Space $X=R^M$

Figure 8.13 The simultaneous probability density of the probability competition neural network

occurrence probability is used for the identification of inputs, and the estimated inference probability is used for classification. The usual multi-layered perceptron estimates only the latter, and thus it cannot identify inputs.

It has been shown that, if the number of hidden units is taken sufficiently large, the multi-layered perceptron can approximate any continuous function by the uniform norm on an arbitrary compact set. We can rigorously prove that, if the number of hidden units is taken sufficiently large, the probability competition neural network can approximate any probability distribution on the direct product of the input space and the output space. Namely, the above neural network can learn any probabilistic relation between the input and the output.

The occurrence probability can be calculated as

$$P(w;x) = \frac{1}{Z(\theta)} \sum_{h=1}^{H} \frac{1}{(2\pi\rho_h^2)^{K/2}} \exp\left(-\frac{\|x-x_h\|^2}{2\rho_h^2} + \theta_h\right)$$

$$= \frac{1}{Z(\theta)} \sum_{h=1}^{H} R_h(x) \tag{8.19}$$

where we use the notation

$$R_h(x) = \frac{1}{(2\pi\rho_h^2)^{K/2}} \exp\left(-\frac{\|x-x_h\|^2}{2\rho_h^2} + \theta_h\right)$$

The expected value $E(w;x)$ of the inference probability is given by

$$E(w; x) = \int yP(w; y|x)dy = \frac{\int yP(w; x, y)dy}{\int P(w; x, y)dy}$$

(8.20)

$$= \frac{\sum_{h=1}^{N} y_h R_h(x)}{\sum_{h=1}^{N} R_h(x)}$$

The estimated occurrence probability and the estimated inference probability can be calculated by the neural network given in Figure 8.14.

$R_h(x)$ is calculated from the input units to the first hidden units. $R_h(x)$ is summed up in the normalizing unit, which is in proportion to the occurrence probability

$$o(x) = \sum_{i=1}^{H} R_h(x) = Z(\theta)P(w; x)$$

(8.21)

By using the output of the normalizing unit, we have

$$E(w; x) = \sum_{i=1}^{H} y_h \frac{R_h(x)}{o(x)}$$

From the first hidden units to the second hidden units, the outputs of the first hidden units are divided by the output of the normalizing unit. From the second hidden units to the output units, the linear sum of y_h is calculated, which is the output of the neural network. By this simple procedure, the estimated occurrence probability and the estimated inference probability can be calculated using parallel processing. The training rule for the probability competition neural network can be introduced based on the maximum likelihood method:

$$\Delta w = -\frac{\partial}{\partial w} l_{emp}(w)$$

where $l_{emp}(w)$ is the empirical loss function,

$$l_{emp}(w) = -\frac{1}{N} \sum_{i=1}^{N} \log P(w; x_i, y_i)$$

(8.22)

To consider the maximum likelihood method in detail, we define the prediction loss function by

$$l(w) = -\int Q(x)\log P(w;x)dx \tag{8.23}$$

If the number of training samples is sufficiently large, then

$$
\begin{aligned}
l_{emp(w)} &\approx l(w) \\
&\le -\int Q(x,y)\log Q(x,y)dxdy
\end{aligned}
$$

where the equality holds if and only if $P(w;x,y) = Q(x,y)$. That is to say, by the maximum likelihood method, the parameters are optimized based on the maximum entropy criterion.

The above training rule results in the following rules by simple calculation:

$$\Delta\theta_h = c_h \sum_{i=1}^{N} \{d_{hi} - 1\}$$

$$\Delta\xi_h = c_h \sum_{i=1}^{N} \frac{d_{hi}}{\rho_h^2} \{x_i - \xi_i\} \tag{8.24}$$

$$\Delta\rho_h = c_h \sum_{i=1}^{N} \frac{d_{hi}}{\rho_h^3} \{\|x_i - \xi_h\|^2 - K\rho_h^2\}$$

$$\Delta\eta_h = c_h \sum_{i=1}^{N} \frac{d_{hi}}{\sigma_h^2} \{y_i - \eta_i\}$$

$$\Delta\sigma_h = c_h \sum_{i=1}^{N} \frac{d_{hi}}{\sigma_h^3} \{\|y_i - \eta_h\|^2 - L\sigma_h^2\}$$

where we used the notation

$$c_h = \frac{\exp(\theta_h)}{Z(\theta)}$$

$$d_{hi} = \frac{1}{P(w;x_i,y_i)} \frac{\exp\left(-\dfrac{\|x_i - x_h\|^2}{2\rho_h^2} - \dfrac{\|y_i - y_h\|^2}{2\sigma_h^2} + \theta_h\right)}{(2\pi\rho_h^2)^{K/2}(2\pi\sigma_h^2)^{L/2}}$$

Inference Probability $P(w;y|x)$

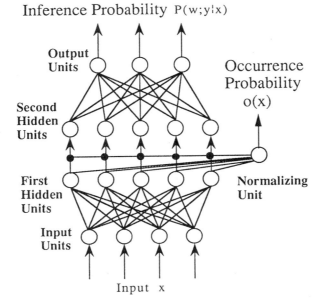

Figure 8.14 A probability competition neural network (PCNN)
(A PCNN can estimate both the occurence probability and the inference probability)

To find the maximum likelihood estimator, we can use the EM algorithm instead of the steepest descent of the parameters [15]. By applying the EM algorithm, the faster learning rule can be obtained.

To find the optimal model, the modified information criterion can also be used in this case, because there is a relation between the empirical loss function and the prediction loss function [4]:

$$< l(w^*) >=< l_{emp}(w^*) > + \frac{(F(w^*)+1)}{N} \qquad (8.25)$$

Therefore, a modified information criterion can be obtained by replacing $F(w)$ by $F_\alpha(w)$. Let us consider a simple example which shows how a PCNN can be applied to information processing. We choose the generalized EXOR problem for $Q(x,y)$,

$$Q(x_1,x_2,y) = \frac{1}{4} \sum_{i=1}^{4} \frac{1}{(2\pi\sigma^2)^{3/2}} \exp\left(-\frac{(x_1-a_{1i})^2 + (x_2-a_{2i})^2 + (y-b_i)^2}{2\sigma^2}\right)$$

where

$$\{(a_{1i}, b_{2i}, c_i); i = 1,2,3,4\} = \{(0,0,0), (1,0,1), (0,1,1), (1,1,0)\}$$

and $\sigma = 0.2$. Figure 8.15 (1) shows training samples independently taken from this probability distribution. A probability competition neural network with four hidden units is trained to approximate this distribution. The initial parameters were set randomly. It took 500 training sessions for the PCNN to learn 100 learning samples. Figure 8.15 (2) shows the estimated regression function by the multi-layered perceptron with four hidden units after 10 000 training sessions, (3) shows the estimated regression function, and (4) shows the estimated occurrence probability (outputs of the normalizing unit). These diagrams show that a probability competition neural network can approximate both the inference probability and the occurrence probability, with the result that it can not only classify an input vector into a category but also identify a given input. Figure 8.15 (2) shows that the multi-layered perceptron is not always appropriate even for estimation of the averaged output.

Since the PCNN estimates the simultaneous probability density, it can also estimate the inverse inference probability $Q(x|y)$. The inverse inference probability has an interesting application. For example, in the character recognition system, the PCNN can illustrate character images from a given character code, based on the inverse inference probability. This fact means that the PCNN can learn the character concepts not only by the usual training but also by dialogue with an operator. Figure 8.15 (5) shows the inverse inference probability for $\{y = 0\}$, and (6) shows the inverse inference probability for $\{y = 1\}$.

8.5 APPLICATION TO ULTRASONIC OBJECT IDENTIFICATION

In this section, we consider a practical application of the probability competition neural network to ultrasonic 3-D object identification. From the 3-D ultrasonic image $f(x,y,z)$ in equation (8.17), we first compute a feature value which is theoretically invariant under parallel shift and rotation.

$$s(r,z) = \int_{D(r)} f(x,y,z)dxdy \tag{8.26}$$

where

$$D(r) = \{(x,y); r^2 \le (x - x_g)^2 + (y - y_g)^2 < (r+a)^2\}$$

and (x_g, y_g) is the gravity centre of $f(x,y,z)$,

$$x_g = \frac{\int xf(x,y,z)dxdydz}{\int f(x,y,z)dxdydz}, \quad y_g = \frac{\int yf(x,y,z)dxdydz}{\int f(x,y,z)dxdydz}$$

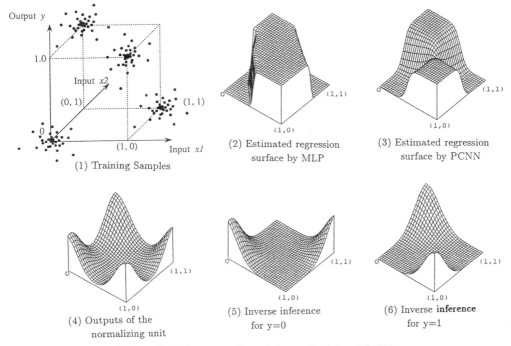

Figure 8.15 An experimental result of the PCNN
(The PCNN can not only classify the inputs into categories but also identify the inputs)

Because of the noises and the finite number of image pixels, even the invariant feature value is fluctuated in the real problem. To develop a machine which can identify such a feature value, we apply the probability competition neural network. Figure 8.16 shows the block diagram of the recognition process. First, the pressures of the scattered waves are observed, and a 3-D image of the object is calculated by ultrasonic imaging. The ultrasonic image is then transformed into the above feature value. The feature value is inputted into the probability competition neural network, and, if it is unknown, then it is rejected, or if it is known then it is classified into some category. Finally, its category is outputted.

In the experiment, thirty objects in Figure 8.11 were used as training samples. Three objects in Figure 8.17 were contained in the training samples, and the other three were not. Two groups of samples were collected:

[Group 1] Thirty objects in Figure 8.11 were placed at the origin and rotated 0 and 45 degrees. Ten samples were collected for each object and rotation angle.

[Group 2] Thirty objects in Figure 8.11 were placed at the place 20 mm from the origin, and rotated 0, 5, 10, 15, 20, ... , 45 degrees. Ten samples were collected for each object and angle.

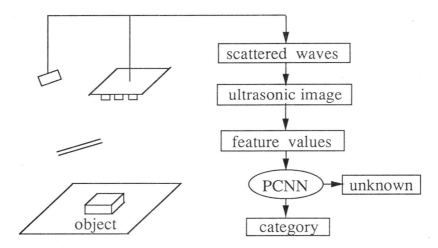

Figure 8.16 Block diagram of the 3-D object identification system
(Unknown objects are rejected and learned objects are classified)

A multi-layered perceptron and a probability competition neural network were trained by using samples in Group 1. Both networks were tested by using samples in Group 2. The recognition rates with the given number of hidden units are shown in Figure 8.18. It is clear that both networks can classify objects at almost the same rate. The probability competition neural network needed more hidden units than the multi-layered perceptron, because the probability competition neural networks estimates both the occurrence probability and the inference probability.

Table 8.1 shows the outputs of the normalizing unit of the probability competition neural network. When learned objects were inputted, outputs were larger, and when unknown inputs were inputted, outputs were smaller. This result shows that the probability competition neural network can reject unknown objects.

Figure 8.17 Learned and unknown objects

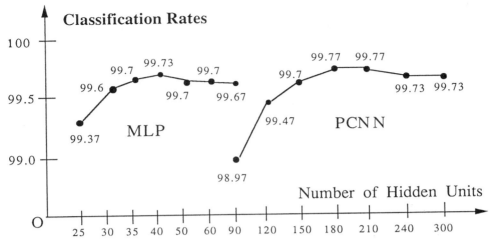

Figure 8.18 Recognition rates by MLP and PCNN

8.6 CONCLUSIONS

In this chapter, we have discussed the statistical properties of artificial neural networks and their application to ultrasonic object recognition. In the first half of the chapter, the three-layered perceptron for ultrasonic image reconstruction was optimized for the minimum prediction error, based on the modified information criterion. In the latter half, the probability competition neural network, which can not only classify inputs into categories but also reject unknown inputs, was trained to identify 3-D objects. We have tried to show that artificial neural networks can be established as part of engineering based on the statistical framework, and that the probabilistic design methods for artificial neural networks are useful in practical applications.

	Objects	$\log o(x)$
Learned Objects	Cube	-5.5
	Block	-12.3
	Spanner	-7.2
Unknown Objects	Sphere	-72.1
	Pyramid	-136.3
	Cylinder	-56.5

Table 8.1 The PCNN can answer 'UNKNOWN'
(The output of the normalizing unit shows how familiar the PCNN is with a given input)

One of the central problems in statistics is how to estimate the environmental probability distribution, and many approaches have been studied: parametric estimation, non-parametric estimation as Parzen's window [16], and semi-parametric estimation. There are many different approaches even in the parametric estimation, for example, Bayes estimation [17], maximum *a posteriori* estimation, and maximum likelihood estimation. Artificial neural networks can be understood as nonlinear parametric models in statistics, and there remain many open problems on the statistical properties of artificial neural networks. Does the maximum likelihood estimator, which minimizes the empirical error, always exist? What distribution is the maximum likelihood estimator subject to, when the Fisher information matrix is degenerate? How do the local minima of the error function affect generalization in learning? Can we rigorously analyze such problems by constructing solvable models [18] [19] ? These are the problems for future study.

Finally, we would like to claim that artificial neural networks now face a serious obstacle. The maximum likelihood method realizes the fastest estimation of the unknown parameters, or, in other words, it approximately achieves the Cramer-Rao lower bound. Therefore, if we can assume that the real environment is subject to a probability distribution, then the statistical training algorithm is the optimal method for artificial neural networks to perceive the real world. However, is the real world, which includes human beings, really subject to some probability distribution? Even if we can assume that the real world is subject to a probability distribution, how can we construct a learning machine which can estimate such an infinitely-high-dimensional distribution with phase transitions? We hope that mathematical research in biological and artificial neural networks will open up a new perspective.

8.7 ACKNOWLEDGEMENTS

The author would like to thank Prof. S. Amari and all members of the Amari seminar at the University of Tokyo for discussions on the statistical foundation of artificial neural networks. Also, the author would like to thank Prof. S. Ueha at the Tokyo Institute of Technology for discussions on ultrasonic imaging.

REFERENCES

[1]   H.White, "Learning in artificial neural networks: a statistical perspective", *Neural Computation*, 1, 425-464 (1989).

[2]   S.Watanabe, K.Fukumizu, "The unified neural network theory and proposal of new models", *Proc. of IJCNN (Beijing)*, 2, 381-386 (1992).

[3]   S.Amari, N.Fujita, S.Shinomoto, "Four types of learning curves", *Neural Computation*, 4, 608-618, (1992).

[4]   H.Akaike, "A new look at the statistical model identification", *IEEE Trans. on Automatic Control*, AC-19, 629-636, (1974).

[5]   S.Watanabe, M.Yoneyama, "An ultrasonic visual sensor for three-dimensional object recognition using neural networks", *IEEE Trans. on Robotics and Automation*, RA-8, 2, 240-249, (1992).

[6]   S.Watanabe, M.Yoneyama, "An ultrasonic visual sensor using a neural network and its application for automatic object recognition", *Proc. of IEEE Ultrasonics Symp.*, 781-874 (1991).

[7]   S.Watanabe, M.Yoneyama, "An ultrasonic 3-D object recognition method based on the unified neural network theory", *Proc. of IEEE Ultrasonics Symp.*, 1191-1194 (1992).

[8]   C.Schwarz, "Estimating the dimension of a model", *Ann. Statist.*, 6, 461-464, (1978)

[9]   J.Rissanen, "Universal coding, information, prediction, and estimation", *IEEE Trans. on Information Theory*, IT-29, 629-636 (1984).

[10]   A.R.Barron, "Universal approximation bounds for superposition's of a sigmoidal function", to appear in *IEEE Trans. on Information Theory*.

[11]   S.Watanabe, "An optimization method of layered neural networks based on the modified information criterion", to appear in *Advances in Neural Information Processing Systems*, 6 (1994).

[12]   S.Watanabe, "An optimization method of artificial neural networks based on a modified information criterion", *IEICE Technical Report*, NC93-52, 71-78 (1993).

[13]   A.S.Weigend, D.E.Rumelhart, and B.A.Huberman, "Generalization of weight-elimination with application to forecasting", *Advances in Neural Information Processing Systems*, 5,598-604 (1992).

[14]   M.D.Richard, R.P.Lipmann, "Neural network classifiers estimate Bayesian a posteriori probabilities", *Neural Computation*, 3, 461-483 (1991).

[15]   A.P.Dempster, N.M.Laird, and D.B.Rubin, "Maximum likelihood from incomplete data via EM algorithm", *Jounal of the Royal Statistical Society*, B, 1-38 (1977).

[16]   E.A.Nadaraya, *Nonparametric estimation of probability densities and regression curves*, Kluwer Academic Publishers (1989).

[17]   E.Levin, N.Tishby, and S.A.Solla, "A statistical approach to learning and generalization in layered neural networks", *Proc. of the IEEE*, 78, 1568-1574 (1990).

[18]   R.J.Baxter, *Exactly solved models in statistical mechanics*, Academic Press (1982).

[19]   S.Watanabe, "Solvable models of artificial neural networks", to appear in *Advances in Neural Information Processing Systems*, 6, (1994).

9

Visual control of robotic manipulator based on neural networks

Takashi Kubota[1] **and Hideki Hashimoto**[2]
[1] Institute of Space and Astronautical Science, Kanagawa, Japan
[2] Institute of Industrial Science, University of Tokyo, Tokyo, Japan

9.1 INTRODUCTION

In recent years, systems which integrate both visual sensors and robots together have received a lot of attention, especially in the field of intelligent robots [1-4]. Such systems can solve many problems which limit the applications of current robots. Using external sensors, a robot can have an *adaptive behaviour*: the robot is able to deal flexibly with changes in its environment and to execute intelligent tasks [5-7].

To make robot control decisions in a task-referenced space, it is necessary to transform sensor outputs into decision space. Several researchers have discussed possibilities for the application of neural networks in robot control [8-11]. The basic theme of such discussion is that of using the network to learn the characteristics of the robot system. Methods to perform inverse kinematics of a multiple joint arm or the inverse dynamic model [12,13] and mappings from the sensory domain to the joint angle or the actuator torque domain have been researched [14-16]. Miller [17-18] proposed a learning control approach that utilizes a CMACS neural network model. In the controller, the network is used as a feedforward term in place of an explicit system model. Kawato *et al.*[12] proposed a hierarchical neural network model that accounts for the generation of motor command. An inverse dynamics model is realized by identifying the parameters of the system whose model is partially known. Pao[19] investigated the potential for the use of artificial neural nets in the control of several systems.

In this chapter a control scheme [20] for a robotic manipulator with visual sensors is presented, which makes use of visual information to position and orientate the end-effector. In the scheme the position and the orientation of the target workpiece with respect to the base frame of the robot are assumed to be unknown, but the desired relative position and orientation of the end-effector to the target workpiece are given as image data in advance. The proposed system controls the robot so that it can approach the desired position and orientation from arbitrary initial ones. The end-effector position and orientation control loop is closed using visual data to generate the necessary control inputs for the manipulator joints by two neural networks. The task considered here is to move the manipulator end-effector into a position where gripping of an object can easily be performed. The relations between the image data of the object and the joint angles of the desired manipulator end-effector position and orientation are clearly nonlinear. The system organizes itself for any manipulator configuration by learning this nonlinear mapping regardless of joint type and geometric dimension. Thereby, the inverse kinematic solution need not be calculated. The control system directly integrates visual data into the servoing process without subdividing the process into determination of the position and orientation of the workpiece and inverse kinematic calculation [21]. A global network learns control signals for larger, and a local network for smaller, object distances. The generalization ability of the neural networks assures control robustness and adaptability in the event of slightly changed object positions [22].

This chapter is structured as follows. In Section 9.2, the general concept of the system is given and explained. Then the control strategy is discussed in Section 9.3. Here the structure, the principle, and the learning process of the artificial neural network are also explained. In Section 9.4, the results of computer simulations are shown and in Section 9.5 experimental results are shown. In Section 9.6, the geometric approach is discussed.

9.2 PROBLEM STATEMENTS

9.2.1 Setting of problem
In robotic application, the task of pick and place is the most fundamental one for robotic manipulators, as shown in Figure 9.1. Usually, this task consists of the following sub-tasks:

1. Guiding a robotic manipulator to a workpiece
2. Letting it pick the workpiece
3. Moving it to another place.

In this study, the first sub-task - guiding a robotic manipulator to a workpiece - is discussed. To pick a workpiece, the position and the orientation of the manipulator play important roles. The manipulator must keep the desired relative position and orientation to the workpiece placed in arbitrary positions. This constrained motion of the manipulator removes most of the difficulties included in other sub-tasks, such as picking a workpiece. If the manipulator is in the desired relative position and orientation to the workpiece after sub-task (1), sub-tasks (2) and (3) can be performed easily, because small position errors can be compensated by the compliance.

Figure 9.2 shows the visual control system discussed in this chapter. The objective is to move the end-effector of a robotic manipulator towards a target point (workpiece). The camera is mounted on the end-effector of the z axis of the end-effector frame with some offset. This chapter discusses the control system that uses visual information to position and orientate the end-effector of the manipulator. The movement of the manipulator is achieved by generating the control signals through the use of an artificial neural network.

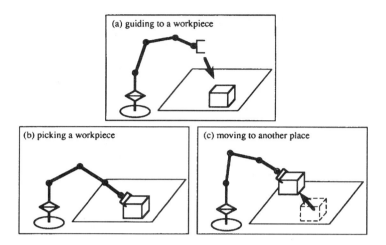

Figure 9.1 Pick and place

9.2.2 Transformation of coordinate system

In general, the coordinate transformation of the following four steps is needed in order to determine the control inputs to the joint angles from visual data. The procedure of coordinate transformation is shown in Figure 9.3.

Step 1: (X_c, Y_c, Z_c) and (ϕ_c, φ_c, ψ_c) the position and the orientation of the camera with respect to the world coordinate system, respectively (representation in Eulerian angles). (X_i^P, Y_i^P, Z_i^P)   is the position of feature point i on the workpiece with respect to the world coordinate system, and (X_i^C, Y_i^C, Z_i^C)   is the position of feature point i on the workpiece with respect to the camera coordinate system. The coordinate transformation of Step 1 is expressed by the rotation matrix R in (9.1):

$$\begin{bmatrix} X_i^C \\ Y_i^C \\ Z_i^C \end{bmatrix} = R^T \begin{bmatrix} X_i^P - X_c \\ Y_i^P - Y_c \\ Z_i^P - Z_c \end{bmatrix} \tag{9.1}$$

Step 2: The image plane (X_i^I, Y_i^I) corresponding to the $X^C Y^C$ plane at distance f (the focal length) from the position of the camera frame to the image plane can be written as

$$\begin{bmatrix} h\,X_i^I \\ h\,Y_i^I \\ h \end{bmatrix} = \begin{bmatrix} f & 0 & 0 \\ 0 & f & 0 \\ 0 & 0 & 1 \end{bmatrix} \begin{bmatrix} X_i^C \\ Y_i^C \\ Z_i^C \end{bmatrix} \tag{9.2}$$

where h is constant.

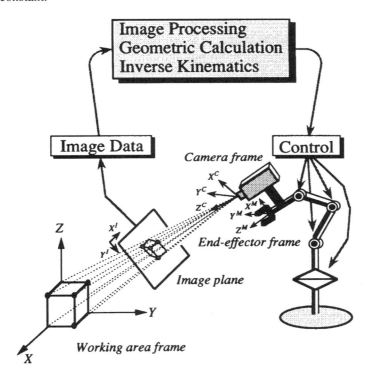

Figure 9.2 Visual control

Step 3: The transformation of Step 3 is the rotation and translation from (X_i^C, Y_i^C, Z_i^C) with respect to the camera coordinate system to (X_i^M, Y_i^M, Z_i^M) with respect to the end-effector coordinate system of the manipulator by using some offset.

Step 4: In Step 4, inverse kinematics can be solved to determine the joint angles of the manipulator from the position (X, Y, Z) and the orientation (ϕ, φ, ψ) of the end-effector of the manipulator.

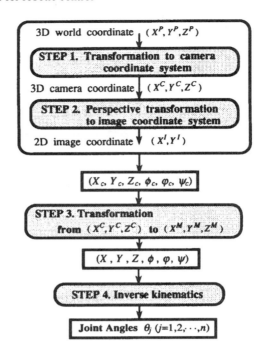

Figure 9.3 Transformation of coordinate system

9.2.3 Decision of position and orientation from visual data

If the position and the orientation of the end-effector of the manipulator are known in advance, it is possible to obtain joint angles by using the coordinate transformation of Step 3 and Step 4. In equations (9.1) and (9.2), the known parameters are the focal length f of the camera lens and the positions of feature points in the image plane (X_i^I, Y_i^I). The unknown parameters are the position (X_c, Y_c, Z_c) and the orientation ($\phi_c, \varphi_c, \psi_c$). However, it is impossible to identify the unknown parameters under these conditions. Therefore, the condition is added into the system that the configuration of the workpiece is known in advance. That is, it is possible to determine the unknown parameters (X_c, Y_c, Z_c) by using the distance (D_{ij}) between the feature points with respect to the world coordinate system. The relation among these parameters can be written as in equations (9.3) and (9.4).

$$X_i^I = f\frac{(C\phi C\varphi C\psi - S\phi S\psi)(X_i^P - X_c) + (S\phi C\varphi C\psi + C\phi S\psi)(Y_i^P - Y_c) + S\phi C\varphi)(Z_i^P - Z_c)}{-C\varphi S\psi(X_i^P - X_c) + S\phi S\psi(Y_i^P - Y_c) + C\varphi(Z_i^P - Z_c)}$$

$$Y_i^I = f\frac{(-C\phi C\varphi S\psi - S\phi C\psi)(X_i^P - X_c) + (-S\phi C\varphi C\psi + C\phi C\psi)(Y_i^P - Y_c) + S\phi S\varphi)(Z_i^P - Z_c)}{C\phi C\varphi(X_i^P - X_c) + S\phi S\varphi(Y_i^P - Y_c) + C\varphi(Z_i^P - Z_c)}$$

with ($i = 1, 2, 3, 4$), and

$$
\begin{aligned}
S\phi &= \sin{(\phi)}, & C\phi &= \cos{(\phi)} \\
S\varphi &= \sin{(\varphi)}, & C\varphi &= \cos{(\varphi)} \\
S\psi &= \sin{(\psi)}, & C\psi &= \cos{(\psi)}
\end{aligned}
\tag{9.3}
$$

$$
\begin{aligned}
D_{12} &= \sqrt{(X_1^P - X_2^P)^2 + (Y_1^P - Y_2^P)^2 + (Z_1^P - Z_2^P)^2} \\
D_{13} &= \sqrt{(X_1^P - X_3^P)^2 + (Y_1^P - Y_3^P)^2 + (Z_1^P - Z_3^P)^2} \\
D_{14} &= \sqrt{(X_1^P - X_2^P)^2 + (Y_1^P - Y_2^P)^2 + (Z_1^P - Z_2^P)^2}
\end{aligned}
\tag{9.4}
$$

However, it requires a large amount of nonlinear calculation to solve (9.3) and (9.4). Most of the recent research pays attention to reconstructing a three-dimensional shape from two-dimensional image data [23], or to determining the relative position and orientation between the camera coordinate system and the workpiece. On the other hand, there is little research that directly determines the joint angles of the manipulator from image data.

In this chapter, a control strategy is proposed to control the manipulator so that it can approach the desired position and orientation without any coordinate transformation (from Step 1 to Step 4), which needs a lot of calculation.

9.3 CONTROL STRATEGY

9.3.1 Theoretical background
It is assumed that there are four visual cues on the target workpiece and that their relation is known in advance. With these data and the x–y coordinate values of the four points in the camera picture, the current position and orientation of the end-effector relative to the workpiece can be uniquely determined by geometric calculations[24][25]. The camera is mounted on the end-effector of the manipulator so that the optical axis is identical to the z-axis of the end-effector frame with some offset. If the manipulator model (link configuration and length) is assumed to be known and the current values of the joint angles can be obtained from shaft encoders, then the position and orientation of the manipulator with respect to the workpiece can be calculated simply by coordinate transformation.

9.3.2 Visual control scheme
The task is to move the end-effector of the robotic manipulator to the desired position and orientation relative to the workpiece. The amount of movement of the manipulator is determined using image data and neural networks.

According to the above theory, the following visual control strategy is introduced[26]. Figure 9.4 shows two camera images: Figure 9.4(a) is the current image when the end-effector is located at the current position and orientation and Figure 9.4(b) is the desired image (end-effector at the desired relative position and orientation). The visual control system computes the error signals in terms of positional coordinates derived from the visual image. The proposed system generates the change in the joint angles so that the coordinates of the four visual cues on the current image coincide with the desired ones $(A \to A^*, B \to B^*, C \to C^*, D \to D^*)$. One way to transform the deviation ΔS in the camera domain to the joint angle displacement vector $\Delta\theta$ is to construct an inverse model of the system (robot and camera). The deviation Δs means the changes Δx on the coordinates of the four visual cues on the camera image. If Δs is known, the inverse model $M(\theta)$ can be used to compute $\Delta\theta$ in

$$\Delta\theta = M(\theta) \cdot \Delta s \qquad\qquad (9.5)$$

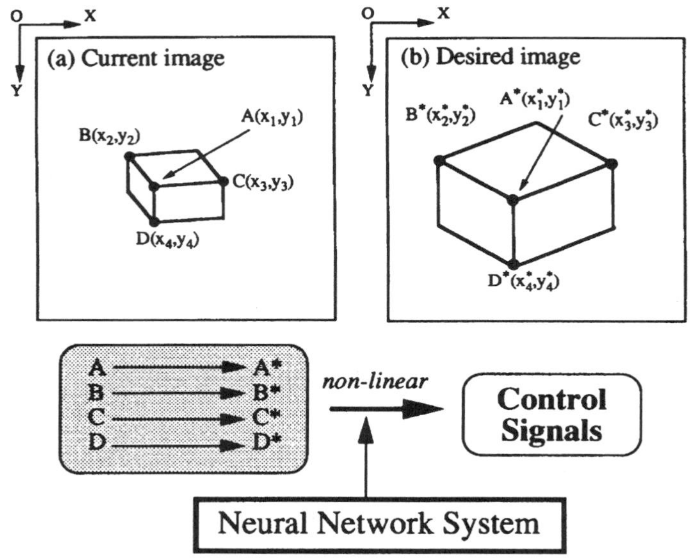

Figure 9.4 Camera image

where the joint angles θ are available. Because this transformation is nonlinear and dependent on θ, it can not be implemented as a constant transformation matrix. Two neural networks are used to learn this nonlinear function. Using conventional image processing techniques, the deviation Δs is determined. Together with the current joint angle vector θ, this positional deviation Δs is used to control the manipulator.

The system is designed to operate under the following conditions. There are many identical workpieces which come into the working area of the robot. A workpiece comes into the visual range of the camera, where its orientation and even position relative to the manipulator are arbitrary. Then the manipulator is required to move its end-effector to the pre-designed position and orientation relative to that workpiece. Here it is assumed that there are four visual cues on the workpiece and that the camera system is able to distinguish these cues from one another.

9.3.3 Neural network system

A PDP network [27] is used as a computational model to learn an unknown mapping from a set of input patterns to a set of output patterns. The input of the neural network consists of the positional vector Δs of the four target cues in the image domain and the current joint angle vector θ. The output of the network gives the required joint angle displacement vector $\Delta\theta$:

$$\Delta\theta = NN(\Delta s, \theta) \qquad (9.6)$$

where NN is the mapping performed by the neural network.

The desired x-y coordinate values of the four visual cues in the camera picture are obtained in advance by moving the end-effector to the desired point. The learning algorithm used in this network is back-propagation [27]. The back-propagation learning algorithm causes a gradient descent on an error function which is the summed square error over all learning samples k:

$$E = \sum_{k} (\Delta\theta_{des}^{k} - \Delta\theta_{act}^{k})(\Delta\theta_{des}^{k} - \Delta\theta_{act}^{k})^{T} \qquad (9.7)$$

where $\Delta\theta_{des}^{k}$ is the kth sampling desired reference to the manipulator and $\Delta\theta_{act}^{k}$ is the kth sampling generated reference by the neural networks. After the neural network learns the relation between input patterns and output patterns sufficiently, it shows a model (function) of the relation between the position and orientation of the end-effector and those of the workpiece. A good model (function) is obtained without using any heuristic knowledge about the manipulator by learning.

The structure of the neural network is essentially suited to parallel processing. The execute time for information processing using the neural network is very fast compared with those using serial processing.

9.3.4 Learning process

The block diagram of the learning process is shown in Figure 9.5. In the learning process, the network learns the nonlinear relation between the joint angles and visual data by the following procedure. First the manipulator is set up in the desired position and orientation to the workpiece. At this stage, Δs and $\Delta\theta$ are zero. Then the random joint angle input $\Delta\theta$ is provided for the manipulator and the visual cues move to different positions in the camera picture. The differences ($\Delta x_i, \Delta y_i$) (i=1,2,3,4) between the positions of the visual cues in the present camera picture and those in the previous picture (desired picture) are obtained. The current joint angle θ with respect to the initial position is sensed by internal sensors. The random joint angle input $\Delta\theta$ is a teaching signal and $\Delta x_i, \Delta y_i$ and θ are input signals to the neural network. A learning pattern PT^k can be defined as :

$$PT^{k} = (\Delta s^{k}, \theta^{k}, \Delta\theta^{k}) \qquad (9.8)$$

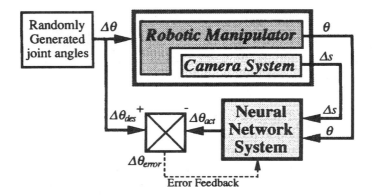

Figure 9.5 Learning process

During the learning process, the weights of connections between neurons are adjusted so that errors (differences between teaching signals $\Delta\theta_{des}$ and outputs $\Delta\theta_{act}$ of the neural network) decrease to zero.

In this process two neural networks are learned. One is the global neural network (NN_G) which is for learning the broad working area of the end-effector (Figure 9.6). Another is the local neural network (NN_L) which is for learning the area in the neighborhood of the workpiece (Figure 9.7). The input of the local neural network is the positional vector Δs of the four target cues in the image domain:

$$\Delta\theta = NN_G \ (\Delta s, \theta \ )$$
$$\Delta\theta = NN_L \ (\Delta s \ ) \tag{9.9}$$

Use of these two neural networks can help reduce the number of training patterns and the number of network connections and can also contribute to efficient learning. Moreover, these networks can cooperate with each other for precise positioning and orientating.

9.3.5 Execution process

In the execution process as shown in Figure 9.8, the networks calculate the required changes in the joint angles in order to set the four cues into the desired positions in the camera picture. In this process the changeover of two neural networks, the global neural network (NN_G) and local neural network (NN_L), is based on the following change rule, as shown in Figure 9.9.

< Change Rule>

If $\alpha_k \leq \alpha_{th}$ (k=1,2,3) and $\Delta x_i, \Delta y_i \leq \beta_{th}$ (i=1,2,3,4) then NN_L is used, otherwise NN_G is used.

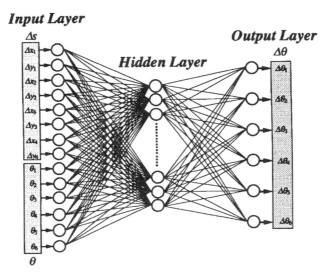

Figure 9.6 Global neural network system

Here,

$$\alpha_1 = \left| \frac{L_{AB}^* - L_{AB}}{L_{AB}^*} \right| , \alpha_2 = \left| \frac{L_{AC}^* - L_{AC}}{L_{AC}^*} \right| , \text{ and } \alpha_3 = \left| \frac{L_{AD}^* - L_{AD}}{L_{AD}^*} \right| \qquad (9.10)$$

L_{mn} is the distance between point m and point n on the current image, L_{mn}^* the distance between point m and point n on the desired image, and α_{th}, β_{th} are threshold values.

After these changes are executed by the manipulator, the camera picture will be checked again and the above process will be repeated if necessary.

9.4 SIMULATION STUDY

Simulation results for a robotic manipulator with six degrees of freedom are shown in this section. A PDP network with three layers is used. The global neural network consists of fourteen input, six output and twenty hidden units (input: Δ s and θ, output $\Delta\theta$). The local neural network consists of eight input, six output and thirty hidden units (input Δ s, output $\Delta\theta$). These nets map the input space, characterizing the four pairs of cues, onto the output space consisting of the six control signals for changes in the manipulator joint angles respectively.

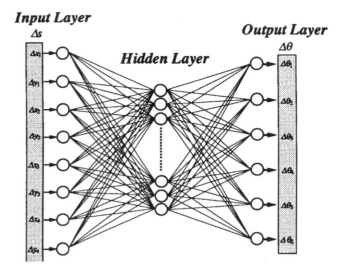

Figure 9.7 Local neural network system

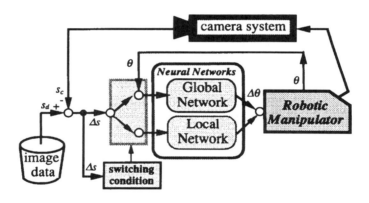

Figure 9.8 Execution process

9.4.1 Learning process

Sixty patterns for the global neural network are produced with the range of angular change as shown in Figure 9.10. Each change is 1.0 degree. One hundred patterns for the local neural network are produced with the range of angular change $-1.5° \leq \theta_i \ (i=1,\cdots,6) \leq 1.5°$ and each change is 0.1 degree. Then, the differences $(\Delta x_i, \Delta y_i)$ (i=1,..,4) between the positions of the visual cues in the present camera picture and those in the previous picture (desired picture) are obtained by moving the manipulator.

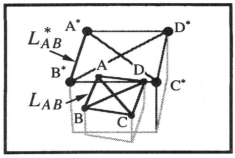

If $\alpha_k \leq \alpha_{th}$ $(k = 1,2,3)$

and $\Delta x_i , \Delta y_i \leq \beta_{th}$ $(i = 1,2,3,4)$

then local neural network,

otherwise global neural network

Figure 9.9 Change rule

Table 9.1 and Table 9.2 show some of these training patterns for the learning processes of the global neural network and the local neural network respectively. These differences are fed to the inputs of the neural network. The errors $\Delta \tilde{\theta}_i$ between the teaching signals $\Delta \theta_i$ and the outputs of the network are decreased to zero by changing the weights of the connections of the neurals by the back-propagation algorithm.

In order to teach the property of the manipulator, the learning process with the back-propagation algorithm was iterated 50,000 times for both the global network and the local network. In this simulation, a NeuralWorks simulator is used.

9.4.2 Execution process

After fifty thousand learning iterations, the pixel errors obtained by moving the manipulator using the global neural network are shown in Figure 9.11 and Table 9.3. Initial errors are in the upper side of each row and final errors after the learning process are in the lower side. It shows that the final pixel errors are reduced to the range -3 to 3. Figure 9.11 shows the simulation results. It is observed that the neural network lets the manipulator approach the workpiece. The end-effector can move to the neighborhood of the desired position and orientation.

9.4.3 Adaptability of neural network

Figure 9.12 shows the adaptability of the neural network. For simulation, the workpiece location is slightly changed as shown in Figure 9.12 (1a). The change of camera images, from (1a) to (1b), is the result of control using the global neural network. In this diagram, positioning errors can be observed, because the location of the workpiece had not yet been learned in the learning process. Here, at this control stage, the network has changed from the global neural network to the local neural network. The result is shown in Figure 9.12(1c). Table 9.4 shows the only small pixel errors.

Table 9.1 Training patterns (global neural network)

Trainig No.	Input Patterns [pixel, °]													
	Δx_1	Δy_1	Δx_2	Δy_2	Δx_3	Δy_3	Δx_4	Δy_4	θ_1	θ_2	θ_3	θ_4	θ_5	θ_6
1	-8	52	28	70	-20	83	23	102	10.0	50.0	13.0	3.0	15.0	5.0
2	-55	18	-25	44	-74	44	-37	72	13.0	50.0	13.0	3.0	13.0	0.0
3	32	-29	67	7	10	3	51	39	8.0	59.0	9.0	0.0	6.0	-2.0
4	-26	71	6	106	-53	102	-13	137	12.0	58.0	14.0	0.0	6.0	-3.0
5	-1	-59	34	-31	-18	-29	24	-1	10.0	54.0	10.0	5.0	8.0	-1.0
6	41	40	71	66	24	68	60	96	8.0	52.0	11.0	4.0	15.0	-4.0
7	-28	8	6	42	-52	40	-11	75	12.0	58.0	12.0	-3.0	5.0	3.0
8	88	-55	115	-19	66	-27	100	9	5.0	56.0	11.0	-4.0	5.0	-3.0
⋮	⋮	⋮	⋮	⋮	⋮	⋮	⋮	⋮	⋮	⋮	⋮	⋮	⋮	⋮
60	55	-66	97	-41	44	-30	92	-5	7.0	57.0	5.0	4.0	11.0	5.0

Trainig No.	Output Patterns [°]					
	$\Delta\theta_1$	$\Delta\theta_2$	$\Delta\theta_3$	$\Delta\theta_4$	$\Delta\theta_5$	$\Delta\theta_6$
1	0.0	-20.0	17.0	-3.0	-5.0	-5.0
2	-3.0	-20.0	17.0	-3.0	-3.0	-5.0
3	2.0	-29.0	21.0	0.0	4.0	2.0
4	-2.0	-28.0	16.0	0.0	4.0	3.0
5	0.0	-24.0	20.0	-5.0	2.0	1.0
6	2.0	-22.0	19.0	-4.0	-5.0	4.0
7	-2.0	-28.0	18.0	3.0	5.0	-3.0
8	5.0	-26.0	19.0	4.0	5.0	3.0
⋮	⋮	⋮	⋮	⋮	⋮	⋮
60	3.0	-27.0	25.0	-4.0	-1.0	-5.0

Table 9.2 Training patterns (local neural network)

Trainig No.	Input Patterns [pixel, °]								Output Patterns [°]					
	Δx_1	Δy_1	Δx_2	Δy_2	Δx_3	Δy_3	Δx_4	Δy_4	$\Delta\theta_1$	$\Delta\theta_2$	$\Delta\theta_3$	$\Delta\theta_4$	$\Delta\theta_5$	$\Delta\theta_6$
1	-39	38	-40	40	-46	37	-45	41	-1.5	-0.4	-0.2	1.0	-1.0	0.7
2	-19	21	-19	22	-22	22	-21	22	-0.7	-0.1	-1.0	1.2	0.7	1.4
3	-11	-14	-12	-13	-12	-16	-12	-15	-0.4	0.4	0.7	1.0	-1.0	-1.2
4	31	32	34	32	33	37	37	36	1.2	-1.2	0.4	1.0	-0.7	-1.3
5	21	6	22	5	22	8	25	7	0.8	-0.7	-0.3	-1.2	1.2	0.9
⋮	⋮	⋮	⋮	⋮	⋮	⋮	⋮	⋮	⋮	⋮	⋮	⋮	⋮	⋮
100	1	3	4	2	3	6	7	4	0.1	-0.5	-0.1	-0.1	0.7	-1.2

	Global Network	Local Network
$\Delta\theta_1$	$-5.0° \sim 5.0°$	$-1.5° \sim 1.5°$
$\Delta\theta_2$	$-30.0° \sim -20.0°$	$-1.5° \sim 1.5°$
$\Delta\theta_3$	$15.0° \sim 25.0°$	$-1.5° \sim 1.5°$
$\Delta\theta_4$	$-5.0° \sim 5.0°$	$-1.5° \sim 1.5°$
$\Delta\theta_5$	$-5.0° \sim 5.0°$	$-1.5° \sim 1.5°$
$\Delta\theta_6$	$-5.0° \sim 5.0°$	$-1.5° \sim 1.5°$

(1.0° step)   ▼   ▼ (0.1° step)

60patterns   100patterns

N.N. $14 \times 20 \times 6$ N.N. $8 \times 30 \times 6$

L.T. 50,000   L.T. 50,000

Figure 9.10 Operating range

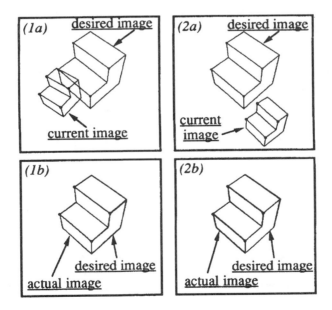

Figure 9.11 Simulation results (learned patterns)

Table 9.3 Pixel errors (learned patterns)

Errors	Δx_1	Δy_1	Δx_2	Δy_2	Δx_3	Δy_3	Δx_4	Δy_4
Initial	-81	31	-52	71	-111	57	-73	99
Final	0	-1	0	-1	-1	-1	0	-1
Initial	58	116	94	142	42	151	84	177
Final	-1	1	-1	2	-2	1	-1	1
Initial	-51	54	-20	86	-76	82	-37	116
Final	1	-3	1	-2	1	-3	2	-1

Table 9.4 Pixel errors (not learned pattern)

Error	Δx_1	Δy_1	Δx_2	Δy_2	Δx_3	Δy_3	Δx_4	Δy_4
Initial	-29	98	1	130	-55	127	-17	160
Global Net	2	25	-2	18	6	23	2	15
Local Net	-3	5	-3	-1	0	5	1	-2
Initial	54	30	96	53	43	68	92	92
Global Net	-10	-22	-4	-15	-14	-19	-6	-12
Local Net	0	-4	4	0	-2	-2	4	2
Initial	-4	89	26	122	-30	118	8	152
Global Net	11	27	4	23	13	23	6	20
Local Net	-8	0	-8	-5	-6	1	-5	-5

Results have shown that the proposed visual control system with neural networks can move the end-effector to the desired location adaptively, even in the case where the location of the workpiece differs slightly from the learned pattern. If an object moves significantly, another method is used to approach the neighborhood of the desired position. And then our scheme is applied to compensate the remaining error.

It took some hours for the neural networks to learn the training patterns on a microcomputer (8086), however, the learning process is off-line. Once neural networks learn the relation between image data and control input, a robotic manipulator can be

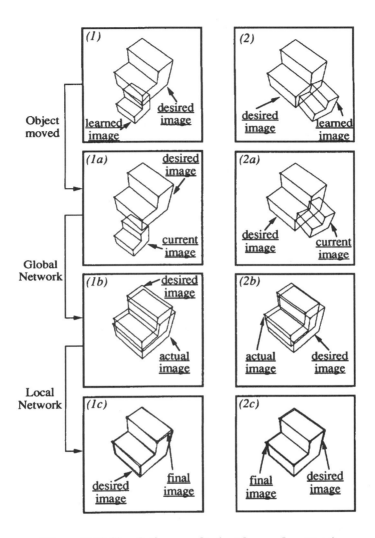

Figure 9.12 Simulation results (not learned pattern)

controlled in real time. Because the structure of the neural network is suited to parallel processing, execution time using neural networks is very fast.

9.4.4 Least squares estimation method

The relation between the image data vector Δs and the joint angular displacement vector $\Delta \theta$ is non-linear and dependent on θ. However, the relation can be regarded as linear if the end-effector is located very close to the desired position. Therefore, the relations between Δs and $\Delta \theta$ are expressed as follows:

$$\Delta\theta = C \, \Delta s$$
$$(\Delta\theta \in R^6, \; C \in R^{6\times8}, \; \Delta s \in R^8 \tag{9.11}$$

The unknown parameter C is estimated by the recursive least squares method. Now the ith joint angular control input is expressed as follows:

$$\Delta\theta_i^k = \Delta s^{kT} c_i + r^k$$
$$c_i^T = [c_{i1}, c_{i2}, \cdots, c_{i8}]$$
$$\Delta s^T = [\Delta s_1, \Delta s_2, \cdots, \Delta s_8] \tag{9.12}$$
$$k = 1, 2, \cdots, N \; \text{(number of data)}$$

Therefore the least squares estimation of the parameter C is given as

$$\widehat{C}_i = (\sum_{t=1}^N \Delta s^k \Delta s^{kT})^{-1} \cdot (\sum_{t=1}^N \Delta s^k \Delta \theta_i^k) \tag{9.13}$$

The recursive algorithm can be formulated as

$$\widehat{C}_N = \widehat{C}_{N-1} + \frac{P_{N-1} \Delta s^N}{1 + \Delta s^{NT} P_{N-1} \Delta s^N} (\theta^N - \Delta s^{NT} \widehat{C}_{N-1}) \tag{9.14}$$

$$P_N = P_{N-1} - \frac{P_{N-1} \Delta s^N \, \Delta s^{NT} P_{N-1}}{1 + \Delta s^{NT} P_{N-1} \Delta s^N} \tag{9.15}$$

The related initial values at the start of the recursive estimation are set as,

$$\widehat{C}_0 = [\,0\,] \; \text{and} \; P_0 = \beta \, I \tag{9.16}$$

where β is a real constant, which should be sufficiently large, such as 10^4 to 10^5.

Table 9.5 and Table 9.6 show the simulation results by the local neural network (LNN) and by least squares estimation (LSE) respectively. In this simulation, N (the number of data) is 100. In the case of least squares estimation, there are some errors for the learned pattern because the sum of the squared errors of all the patterns is minimized. The estimation method is valid only within the neighborhood of the desired position. Therefore, the neural network approach is better than the estimation method.

9.5 EXPERIMENTAL STUDY

9.5.1 Experimental system

An overview of the experimental system is shown in Figure 9.13. A robotic manipulator with five degrees of freedom is used in the experiment. A color CCD (coupled charged

Table 9.5 Pixel errors by neural network method

Error	Δx_1	Δy_1	Δx_2	Δy_2	Δx_3	Δy_3	Δx_4	Δy_4
Initial	-41	14	-38	16	-29	28	-26	7
Final	-1	2	-1	0	-1	0	-2	1
Initial	-19	-12	-32	-13	-41	-23	-40	1
Final	-2	2	-4	-1	-5	1	-4	-1
Initial	28	-35	27	-38	20	-45	18	-30
Final	2	-3	1	-4	2	-7	3	-5

Table 9.6 Pixel errors by least square method

Error	Δx_1	Δy_1	Δx_2	Δy_2	Δx_3	Δy_3	Δx_4	Δy_4
Initial	-41	14	-38	16	-29	28	-26	7
Final	-7	2	-5	3	-4	5	-8	2
Initial	-19	-12	-32	-13	-41	-23	-40	1
Final	-5	4	-9	-6	-12	7	-5	-2
Initial	28	-35	27	-38	20	-45	18	-30
Final	4	-10	6	-6	4	-11	6	-9

device) camera is mounted on the end-effector. Figure 9.14 shows the system configuration. The robot is driven by DC motors. The image obtained from the CCD camera is converted into image data (resolution 256*256, intensity level 6 bits) through the image memory board [FDM98RGB]. The image data are transported by DMA to the VRAM of the microcomputer within about 0.03 s. Trp-Simulator (Ami), which consists of five transputers, is used as a neural simulator, as shown in Figure 9.15.

9.5.2 Neural networks

The robot is moved about the operating range as shown in Figure 9.16 and Figure 9.17 because of range of vision. Sixty learning patterns are given to the global neural network and 100 learning patterns are given to the local neural network. The number of learning times is about 50,000.

9.5.3 Experimental results

Firstly experiments for learned pattern data are done. Experimental results show that the final pixel errors are reduced to the range -5 to +5 as shown in Table 9.7. The end-effector can move to the neighborhood of the desired position and orientation.

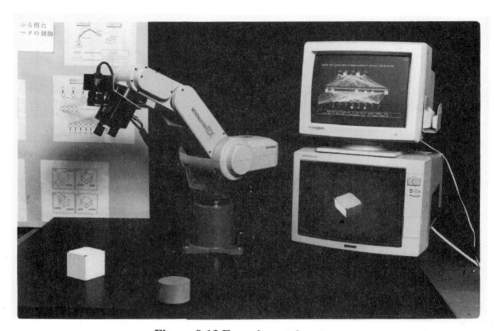

Figure 9.13 Experimental system

Table 9.7 Experimental results (learned patterns)

Pixel errors	Δx_1	Δy_1	Δx_2	Δy_2	Δx_3	Δy_3	Δx_4	Δy_4
Initial	42	-72	-18	-9	41	-9	-20	-71
Final	0	0	-2	-3	-2	0	1	-2
Initial	91	21	37	97	100	87	28	34
Final	4	-3	4	-5	4	-3	4	-5
Initial	3	-3	-59	60	1	58	-60	-3
Final	0	3	1	4	1	3	0	4

One of the requirements of the system is to adapt changes in the environment as shown in Figure 9.18. The manipulator must keep the desired relative position and orientation to the workpiece placed in arbitrary positions. Several experimental results for not-learned pattern data are shown in Table 9.8. Though the current picture is not a learned one, the neural network lets the manipulator approach the workpiece with the desired position and orientation. The final pixel errors for the not learned patterns are reduced to the range -6 to +6. The model learned in the neural network seems to be a correct model of the relative function between the end-effector and the workpiece.

9.6 GEOMETRIC APPROACH

9.6.1 Control strategy

Where the structure of the manipulator and the geometrical position of the features on the object are exactly known, the position and orientation of the manipulator and the object can be calculated by a geometric method [23]. However, the accuracy of the geometric method largely depends upon camera and image memory resolutions and lens linearity, but also depends upon camera system parameters, such as focal length and image center offsets. Therefore, the neural network approach can be used for adjusting position misalignment [28]. The proposed visual control system consists of two mechanisms as shown in Figure 9.19. One is a feedforward control system [29] based on the camera system, robot model, and object model. Another is a feedback control system based on the local neural network.

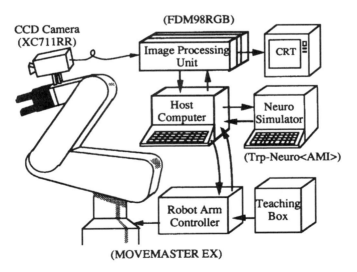

Figure 9.14 Configuration of experimental system

9.6.2 Feedforward controller

In the geometric control approach, the vision system is used as a recognition observer to measure the relative position between the robot end-effector and objects in its environment. This measurement process can be decomposed into some nonlinear transformations. The coordinate transformations of the following six steps are needed in order to determine the control inputs to the joint angles from visual data.

Table 9.8 Experimental results (not learned pattern)

Pixel errors	Δx_1	Δy_1	Δx_2	Δy_2	Δx_3	Δy_3	Δx_4	Δy_4
Initial	3	-69	-53	-14	1	-12	-54	-71
Global Net	-11	-12	-9	-9	-8	-11	-11	-11
Local Net	0	-6	-2	0	1	-3	-2	-3
Initial	37	-63	-17	-9	37	-9	-17	-63
Global Net	12	-12	16	-12	13	-14	16	11
Local Net	3	-3	4	-3	4	-3	4	-2
Initial	0	-42	-53	10	1	12	-54	-44
Global Net	-21	12	-11	7	-17	6	-14	13
Local Net	0	6	1	0	-2	3	3	4

Step 1   The vision system looks at an object and estimates the relative object position P^C_{object}. The transformation matrix from the camera frame to the object frame is represented in terms of the homogeneous coordinate $H_{vloc.act}$.

Step 2   The position and orientation $P^W_{end-effector}$ of the end-effector with respect to the world coordinate system is calculated by kinematics and current joint angles θ_{act}. The position of the end-effector can be represented by the homogeneous coordinate $H_{end-effector.act}$.

Step 3   The position and orientation P^C_{object} of the object with respect to the world coordinate are obtained by Step 1 and Step 2. The homogeneous coordinate H_{object} of the object is calculated as follows.

$$H_{object} = H_{end-effector.act} \cdot H_{trce} \cdot H_{vloc.act} \qquad (9.17)$$

where H_{trce} is the transformation matrix from the end-effector coordinate system of the manipulator to the camera coordinate system.

Step 4 The desired position and orientation P_d of the end-effector with respect to the world coordinate system is calculated as follows:

$$H_{end\text{-}effector.des} = H_{object} \cdot H_{vloc.des}^{-1} \cdot H_{trce}^{-1} \tag{9.18}$$

where $H_{vloc.de}$ is the desired relative position of the object in the camera frame, obtained from the desired image.

Step 5 Inverse kinematics is solved to determine the joint angles θ_d of the manipulator from the desired position (X_d, Y_d, Z_d) and the desired orientation $(\phi_d, \varphi_d, \psi_d)$ of the end-effector of the manipulator.

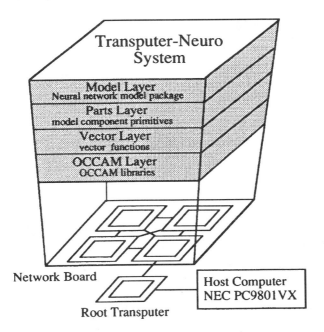

Model : Backpropagation

Network cofiguration :   Input Layer   13unit
                          Hidden Layer   30unit
                          Output Layer   5unit

Learning patterns :   60

Learning times   :   50,000

Learning time   :   10 [hr]

Figure 9.15 Trp-neuro simulator (AMI)

Step 6 Control $\Delta\theta$ to the manipulator is generated by computing the difference between the desired joint angles θ_d and the current joint angles θ_{act}.

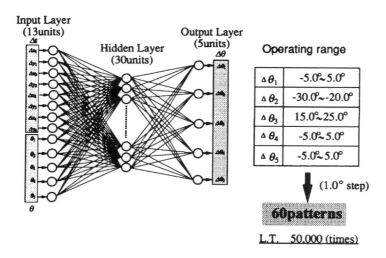

Figure 9.16 Global neural network

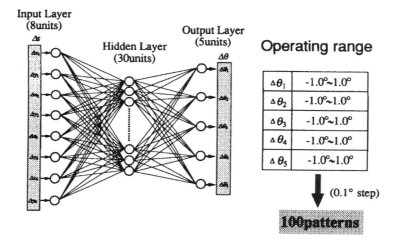

Figure 9.17 Local neural network

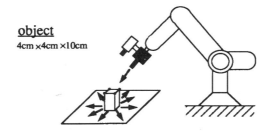

Figure 9.18 Experiment for not learned pattern

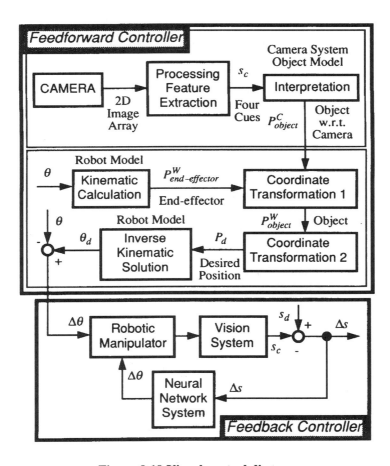

Figure 9.19 Visual control diagram

9.6.3 Feedback controller

A feedback loop is closed by using the neural network to compensate for the small image error. A neural network learns the area in the neighborhood of the workpiece for adjusting workpiece position misalignment. The camera picture is checked again and the feedback loop is repeated until a specified accuracy is achieved. If the visual feedback system is structured so that position estimates and position errors are updated as fast as they are measured, and position corrections are commanded to the manipulator while it is moving, then the dynamic visual servo control system is synthesized by using this approach.

9.7 CONCLUSIONS

In this chapter, a scheme for the control of a robot manipulator with a visual sensor is described. The nonlinear relation between the image data and the control signals for the changes in the joint angles can be learned by artificial neural networks. The validity of this control scheme is confirmed by computer simulations and experimental results. The system organizes itself for a manipulator configuration through a learning process. This approach is effective because it essentially decomposes complex geometric calculations into a simple mapping of the network.

REFERENCES

[1]   R.L.Anderson, "Real time intelligent visual control of a robot," *Proc. of IEEE Int. Workshop on Intelligent Control*, pp.89-94 (1985).

[2]   L.E.Wess, A.C.Sanderson, C.P.Neuman, "Dynamic sensor-based control of robots with visual feedback," *IEEE Journal of Robotics and Automation*, Vol.RA-3, No.5, pp.404-417 (1987).

[3]   Feddema, "Automatic selection of image features for visual servoing of a robotic manipulator," *Proc. of IEEE Int. Conf. on Robotics and Automation*, pp.832-837 (1989).

[4]   P.I.Corke, "Visual control of robot manipulators - a review," *Visual Servoing World Scientific*, pp.1-31 (1993).

[5]   A.C.Sanderson, "Vision and multisensor feedback controllers," *Proc. of IEEE Int. Workshop on Intelligent Control*, pp.80-81 (1985).

[6]   F.Harashima, H.Hashimoto, T.Kubota, "Sensor based robot control systems," *Proc. of IEEE Int. Workshop on Intelligent Motion Control*, Vol.1, pp.1-10 (1990).

[7]   N.Maru, H.Kase *et al.*, "Manipulator control by visual servoing with the stereo vision," *Proc. of Int. Conf. on Intelligent Robots and Systems*, pp.1865-1870 (1993).

[8]   H.Noda, H.Hashimoto, F.Harashima, "A method to construct self-organizing system in robotic application," *Proc. of KACC'88*, pp.1022-1027 (1988).

[9]   T.M.Martinez, H.J.Ritter, K.J.Schulten, "Three-dimensional neural net for learning visuomotor coordination of a robot arm," *IEEE Trans. on Neural Networks*, Vol.1,

No.1, pp.131-136 (1990).

[10]   B.J.A.Krose, M.J.Korst, F.C.A.Groen, "Learning strategies for a vision based neural controller for a robot arm," *Proc. of the IEEE Int. Workshop on Intelligent Motion Control*, Vol.1, pp.1-10 (1990).

[11]   F.J.Sharifi, H.H.Fakhry, W.J.Wilson, "Integration of a robust trajectory planner with a feedforward neural controller for robotic manipulators," *Proc. of IEEE Int. Conf. on Robotics and Automation*, pp.3192-3197 (1994).

[12]   M.Kawato, K.Furukawa, R.Suzuki, "A hierarchical neural network model for control and learning of voluntary movement," *Biological Cybernetics* 57, pp.169-185 (1987).

[13]   A.Guez, Z.Ahmad, "Solution to the inverse kinematics problem in robotics by neural networks," *Proc. of Int. Conf. on Neural Networks* (1988).

[14]   M.Kuperstein, "Neural model of adaptive hand-eye coordination for single postures," *Science*, Vol.239, pp.1308-1311 (1988).

[15]   M.Kuperstein, J.Rubinstein, "Implementation of an adaptive neural controller for sensory-motor coordination," *IEEE Control Systems Magazine* Vol.3, pp.25-30 (1989).

[16]   A.Y.Zomaya, "Trends in neuro-adaptive control for robot manipulators," *Proc. of Int. Conf. on Intelligent Robots and Systems*, pp.754-760 (1993).

[17]   W.T.Miller, "Sensor based control of robotic manipulators using a general learning algorithm," *IEEE J. of Robotics Automat.*, Vol.RA-3, pp.157-165 (1987).

[18]   W.T.Miller III, "Real-time application of neural networks for sensor-based control of robots with vision," *IEEE Trans. on Systems, Man, and Cybernetics*, Vol.19, No.4, pp.825-831 (1989).

[19]   Y.H.Pao, "A connectionist-net approach to autonomous machine learning of effective process control strategies," *Proc. of Int. Conf. Manufacturing Science Technol. Future* (1987).

[20]   H.Hashimoto, T.Kubota, M.Sato, F.Harashima, "Visual control of robotic manipulator based on neural networks," *IEEE Trans. on Industrial Electronics*, Vol.39, No.6 (1992) .

[21]   H.Hashimoto, T.Kubota, M.Baeg, F.Harashima, "A scheme for visual tracking of robot manipulator using neural network," *Proc. of Int. Joint Conf. on Neural Networks*, pp.773-778 (1991).

[22]   H.Hashimoto, T.Kubota, F.Harashima, "Visual control of a robotic manipulator using neural networks," *Proc. of the 29th IEEE Conf. on Decision and Control*, pp.3295-3302 (1990).

[23]   T.Horiguchi, "A 3D visual ranging method using a single camera and its application to an experimental micro-assembly robot," *Proc. of Japan-U.S.A. Symp. on Flexible Automation*, pp.715-720 (1986).

[24]   R.Tsai, T.Huang, "Uniqueness and estimation of three dimensional motion parameters of rigid objects with curved surfaces". *IEEE Trans. on Pattern Analysis and Machine Intelligence*, Vol. 6 (1984).

[25]   Y.Hung, P.Yeh, D.Harwood, "Passive ranging to known planar point sets," *Proc. of IEEE Int. Conf. on Robotics and Automation*, pp.80-84 (1985).

[26]   H.Hashimoto, T.Kubota, M.Sato, F.Harashima, "Visual servo control of robotic manipulators based on artificial neural network," *Proc. of IEEE Int. Conf. on Industrial Electronics*, pp.770-774 (1989).

[27]   D.E.Rumelhart, J.L.McClelland, PDP Research Group, *Parallel distributed* Processing, Bradford Books, Vol.1&2, MIT Press (1986).

[28]   H.Hashimoto, T.Kubota, M.Kudou, F.Harashima, "Self-organizing visual servo system based on neural networks," *Proc. of American Control Conf.*, pp.2262-2267 (1991).

[29]   H.Hashimoto, T.Kubota, M.Kudo, F.Harashima, "Self-organizing visual servo system based on neural networks," *IEEE Control Systems Magazine*, Vol.12, No.4, pp.31-36 (1992).

10

Brain building for a biological robot

Hugo de Garis
Brain Builder Group, Evolutionary Systems Department, ATR Human Information Processing Research Laboratories, 2-2 Hikaridai, Seika-cho, Soraku-gun, Kansai Science City, Kyoto, 619-02, Japan

10.1 INTRODUCTION

This short chapter reports briefly on progress made in the first year of an eight-year research project which aims to build/grow/evolve (at electronic speeds) an artificial brain which contains a billion neurons inside a cellular automata machine (CAM). The chapter introduces the CAM-Brain concept, discusses some implementation issues, and justifies how it should be possible to build artificial brains containing a billion neurons by the year 2001. If successful, the CAM-Brain research project should provide a powerful, if not revolutionary, new tool for the study of behavior, because the project intends to evolve a CAM-Brain which is capable of controlling roughly 1000 behaviors in a robot kitten. Since it is unlikely that all these behaviors will be evolved in one step, the CAM-Brain project will need to focus its attention upon the issue of incremental evolution, i.e. how does one add neural modules to an already functional nervous system, so as to increase its functionality?

The primary motivation behind the idea of building an artificial brain is the realization that progress in micro and nano electronics over the next twenty years will make this possible. To back up this claim, this chapter shows that it will be possible to build/grow/evolve an artificial brain containing a billion neurons by the year 2001. The complexities of such an artificial nervous system will be so large that it will be totally impractical to use traditional engineering design techniques to build it. There will be too many components (e.g. artificial neurons, axons, dendrites, synapses etc.) for the behavior of the total system to be predictable, or even analyzable. Therefore, in order to build such an artificial brain, an evolutionary engineering approach will be used which the author calls genetic programming (GP), i.e. using evolutionary algorithms such as the genetic algorithm [6] as tools to build/grow/evolve complex systems [1,2,3,4,8]. Having decided to use GP to build an artificial brain, the next step was to decide to perform the evolution at electronic speeds in order to accelerate the whole process. This led to the concept of the Darwin machine [8], which is special hardware which performs GP in parallel. However, the direct rewriting of electronic circuitry billions of times (in such devices as PLDs = programmable logic devices) is beyond the state of the art. Therefore, if one wants to rewrite a device an indefinite number of times, one needs to use RAM based systems. This type of thinking led to the idea of using RAM memories as a basis for a hardware implementation of cellular automata, and to grow/evolve neural networks based on cellular automata. Since the Information Mechanics group at MIT has already developed cellular automata machines (CAMs) [7], it seemed appropriate to grow/evolve this artificial brain inside a CAM.

The CAM-Brain Project introduces a new subfield into neural network research, called neurite networks, where the distinction between neural- and neurite- networks is that with the latter, the neural network gets GROWN, i.e. it has an embryological component. A neurite is a neurobiological term meaning a baby neuron which grows connections with other neurites. These artificial neurite networks are based on cellular automata (CA) networks whose branchings are genetically programmed (i.e. they are grown under the control of a genetic algorithm). A sequence of CA signals is sent down the middle of a CA trail (see Figure 10.2). When a signal hits the end of a trail, it makes the trail extend, turn left, turn right, branch left, branch right, split, etc., depending upon the state of the CA signal. These signal sequences are treated as the chromosomes of a Genetic Algorithm. Once the CA network is formed, other CA state transition rules make it behave like a neural network. The fitness of this CA based neural network is measured in terms of how well it performs some task, e.g. controlling some behavior of a biological robot (biot). The Brain Builder Group (a part of the Evolutionary Systems Department) hopes to use such ideas to build Darwin machines (i.e. machines which evolve), based on cellular automata machines, as a tool to build an artificial brain.

This chapter shows how artificial neural networks based on cellular automata can be grown using GP techniques. The ideas and results of this project will serve as the conceptual basis for the construction of Darwin machines. For example, cellular automata machines could function in parallel to evolve the neurite networks described below. Each CAM would have a conventional programmable processor to measure the fitness of the evolved neurite network. A central processor could then perform the genetic algorithm (GA) aspects of the evolution (e.g. calculate the next generation of chromosomes etc.). Alternatively, a more distributed GA could be performed, where each CAM and its processor communicates only with its neighbors. We hope that by using these Darwin machines it will be possible to build/evolve/GP a large number of neurite network modules and their connections to build an artificial brain capable of giving a biological robot (biot) some 1000 behaviors.

This chapter consists of the following sections. Section 10.2 gives a brief introduction to cellular automata and how cellular automata trails can be evolved into cellular automata networks. Section 10.3 expands on the initial ideas of Section 10.2, especially in explaining how CA trails can be made to behave like neural networks. Section 10.4 discusses some of the implementation issues involved in the first year of work.

Section 10.5 shows that it will be possible to build a billion neuron CAM-Brain by the year 2001. Section 10.6 presents ideas for future research in the remaining years of the project. Section 10.7 summarizes the work achieved so far.

10.2 THE GENETIC PROGRAMMING OF CELLULAR AUTOMATA

10.2.1 Automata trails

Cellular automata are cells (e.g. squares in a 2D grid, or cubes in a 3D grid) each of which has one of a finite number of states. State transition rules (applying to all cells in the grid) determine how a cell updates (synchronously) its state depending upon its present state and the states of its neighbors. Figure 10.1 shows an example of a CA state transition rule.

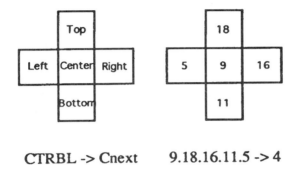

$$CTRBL \rightarrow Cnext \qquad 9.18.16.11.5 \rightarrow 4$$

Figure 10.1 A cellular automata state transition rule

A 3-cell wide CA trail as shown in Figure 10.2 can be fed a sequence of CA signals which propagate down the middle of the trail until they hit the end. When they do, CA state transition rules are defined so that the trail is extended by one square, or made to turn left, turn right, split, branch left, branch right etc. (e.g. [5]). The sequence of these CA signals is then evolved using a conventional genetic algorithm [6]. When one trail collides with another, a synapse is formed, as shown in Figure 10.3. The two cells of the synapse then absorb oncoming signals, thus keeping the configuration of the intersecting trails intact. Figures 10.5 and 10.6 at the end of this chapter show the results of a CA network evolution. In Figures 10.5 and 10.6, there are sixteen CA neurons. The chromosome was split into thirty-two and fed simultaneously into the starting points of the thirty-two CA trails.

10.3 CELLULAR AUTOMATA BASED NEURITE NETWORKS

One can, then, evolve CA networks. There is too little space in this short chapter to go

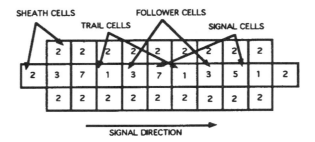

Figure 10.2 A cellular automata trail

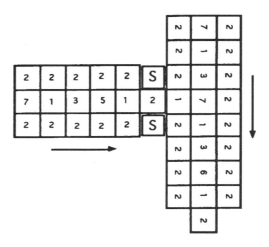

Figure 10.3 A synapse between two CA trails

into the many details, but CA rules can be defined to repair and clean up destroyed trails, i.e. those for which collision circumstances make synapse formation impossible, in which case no CA rules are defined, so by default the background state (black, zero) becomes the next state, which can destroy the trail. Usually the network stabilizes after several hundred clock cycles, i.e. all signal sequences get absorbed at synapses. Once this happens, other CA rules make the CA network behave like a neural network. For example, there are three kinds of sheath cells, two for axons (excitatory and inhibitory) and one for dendrites. Signal strengths in axons keep the same value they had at emission (at CA neurons), but once the axon signal passes through an axon-to-dendrite (A→D) synapse (created in the CA net growth phase), it becomes a dendrite signal, which drops off in strength as it advances. Thus the dendrite signal strength depends on its distance from its (A→D) synapse. Since these distances are evolvable, they are equivalent to the weights of conventional neural networks. Signal values can be positive or negative. At an excitatory synapse, the sign of the axon signal value is transmitted unchanged to the dendrite. At an inhibitory synapse, the sign of the axon signal value transmitted to the dendrite is reversed. Excitatory and inhibitory axons generate excitatory (+ve A→D) and inhibitory (-ve A→D) synapses when dendrite CA trails collide with them. Axon-axon (A→A), and dendrite-dendrite (D→D) synapses are simply not formed. Two merging dendrite signals add their incoming signal strengths at the junction. CA rules can be defined which allow this. When a dendrite CA trail splits, special gating cells are formed at the split junction, which are later used (when the CA trails behave like neural nets) to direct the dendrite signals to turn towards the neuron which grew them. Finally, the axon output signal strength at a CA neuron can be a nonlinear function of the sum of its incoming dendritic signal strengths. The strength of this axon signal remains unchanged as it travels through the axon.

Having introduced the CAM-Brain project, the following two sections will be devoted to issues concerned with CAM-Brain's implementation and performance-scaling (i.e. how quickly could CAM-Brain evolve neural circuits with large numbers of neurons).

10.4 IMPLEMENTATION ISSUES

This section discusses some of the implementation issues resulting from the first year's experience of building CAM-Brain. The research strategy for the next year or so is as follows. Initially a 2D version will be simulated in software on a workstation (Sparc 10). If the 2D version proves to be evolvable, then a 3D software simulation will be undertaken. If the 3D version is evolvable, then work on a modified CAM8 machine [7] will be undertaken. Parallel with this work will be a software simulation of a molecular (nano) scale CAM design. Since this project runs until 2001, we have the luxury to think long term and to anticipate developments in nanotechnology [9] which might be usable in the CAM-Brain project (a nanoCAM-Brain).

The remainder of this section will be concerned with issues that arise during the implementation of the 2D software simulation.

10.4.1 The need for productivity tools
At the time of writing, the author has already hand coded over 7000 cellular automata state transition rules for the 2D version. It is likely that more than 10,000 rules will be needed before the system is fully functional. With such a large number of rules, it is vital that their creation be as efficient as possible, otherwise one can waste many man-months. If other researchers are interested in evolving cellular automata based structures (e.g. neural nets), then, they are advised to team up with a competent programmer who can provide pull down menu windows which allow the following functionalities.

(i) Save and recall screens. It is a waste of time, to have to grow a structure from scratch, just to reach a particular configuration at which a problem occurs. By saving a screen, and being able to recall it easily, one can jump directly to a given configuration.

(ii) Save states. By switching a flag, all the states of the CA can be saved for the latest N cycles. This slows down the growth or signaling process, but enables the   following function.

(iii) Step back. If a problem occurs (e.g. an unwanted black square appears which indicates a missing CA rule) at cycle T, then if the save states flag is on, one can step back (i.e. display the screen at progressively earlier cycles).

(iv) Cell status. By clicking on the mouse on a particular CA cell, its coordinates and state number (color) appear in a movable window.

(v) Cell update. By clicking on an undesired black square (i.e. a CA rule is missing), and typing in the desired state number, an automatic CA state transition rule learning process becomes possible.

(vi) Automatic rule learning. By updating the cell as in (v) above, the software can look at the state of the cell in the previous cycle and the states of its neighbors, and hence can deduce the desired state transition rule and automatically insert the new rule into the rule table.

(vii) Rule consistency check. Every time a rule is inserted (either manually or automatically) a check is made to see whether it is consistent with existing rules, i.e. if the LHS of the new rule (and its rotations) is the same as an already existing rule, then the RHS should be the same. If not, a warning is issued, if so, an already in rule base message appears.

10.4.2 Robustness

One of the main reasons why CAM-Brain has so many handcrafted CA state transition rules is due to the need for robustness. To explain what is meant by this term, a little background information is needed. When one watches many CA networks evolve, one soon realizes that there are many many ways in which CA trails can collide. It is probably not possible to handcode rules for all these collision possibilities.

When a collision occurs for which the necessary CA rules do not exist in the rule base, the next states default to state 0, i.e. black. These black square (cubes) then eat away the CA trail. It is possible to cut off this rot at the nearest synapse, by using repair rules which seal off the decaying trail, but this process destroys (removes) the synapse, and thus reduces the connectivity of the overall CA network, which reduces its evolvability. Therefore, one needs to put in enough rules so that enough of the more common collision possibilities have rules which govern them. The more obscure, i.e. less frequent, collision possibilities do not have rules which govern them. In practice, one finds oneself debugging increasingly improbable collision cases, until nearly all the CA networks are robust enough (i.e. do not get eaten away) to be functional and evolvable. The fact that 2D-CAM-Brain will probably take 10,000 rules to get it to work, obviously implies a considerable investment of human development time. An analogy could be made with the time needed to make a substantial expert system.

10.4.3 Evolvability

The big unanswered question at the moment is whether, after all the 10,000 rule effort is completed, these CA based neural networks will prove to be evolvable, i.e. will their fitness levels continue to increase to the point where their functionality reaches desired levels? The author has already experimented with GenNets [1,2,3,4], i.e. fully connected software simulated neural nets, whose $N*N$ weights (for an N neuron module) were concatenated onto a GA chromosome and evolved.

These GenNets proved to be highly evolvable, as well as being robust. In fact, one could often cut (i.e. make zero) 70% of the weights, and still preserve the evolved functionality. Therefore,it is hoped that the CAM-Brain neural circuits, will also prove to be evolvable, despite their lower connectivity (i.e. lower number of synapses per neuron). Actually, this assumption of lower connectivity will probably not apply to the 3D version. The 2D version of course confines the networks, but with the 3D version (whose implementation has not yet begun), it is likely that axons and dendrites will be able to travel relatively large distances before forming synapses. With these distances and multiple branching, it is probable that the connectivity will be high. Intuitively, one feels that the higher the connectivity, the higher the evolvability, because the solution space in which one is evolving is probably dense, with many different solutions to the same desired evolutionary functionality.

10.5 PERFORMANCE SCALING ISSUES

This section discusses some of the issues which will become apparent when CAM-Brain hopes to scale up to millions even billions of neural network modules. Initially, the time complexity of the current 2D implementation of CAM-Brain on a Sparc10 is presented, and predictions made as to the likely performance of a 3D version using the type of hardware which will be available by the year 2001, the last year of the current CAM-Brain research project.

The current 2D version, is being implemented on a Sun Sparc 10 workstation. It takes approximately 3.4 minutes to grow a stable cellular automata network consisting of only four neurons. It takes a further 3.2 minutes to perform the signaling on the grown network, i.e. a total growth-signaling time to measure the fitness of a chromosome of 6.6 minutes. This time scales linearly with the number of artificial neurons in the network. If one uses a population of ten chromosomes, for 100 generations, the total evolution time (on a Sparc 10) is 100×10×6.6 minutes, i.e. 110 hours, or 4.6 days. This is obviously tediously slow, hence the need to use hardware. As made explicit in the title of this research project, it is the intention to use special hardware, called cellular automata machines (CAMs) to accelerate the evolution. State of the art is MIT's CAM8 [7] which can update 25 million cellular automata cells per second, per hardware module. A CAM8 box (i.e. a personal computer size box) contains eight such modules, and costs about $10,000. Such boxes can be connected blockwise indefinitely, with a linear increase in processing capacity. Assuming an eight module box, how quickly could the above evolution (i.e. 100 generations, with a population size of 10) be performed? With eight modules, 200 million cell updates per second is possible. If one assumes that the 2D CA space in which the evolution takes place is a square of 100 cells on a side, i.e. 10,000 cells, then all these cells can be (sequentially) updated by the CAM8 box in 50 microseconds.

Assuming 1000 CA clock cycles for the growth and signaling, it will take 50 milliseconds to grow and measure the fitness of one chromosome. With a population of 10 and 100 generations, total CAM8 evolution time for a four neuron network will be 50 seconds, i.e. about 1 minute, which is roughly 8000 times faster. Into the same CAM8 box, and a 3D space of a million cells, i.e. a cube of side 100 cells, one could put roughly 40 neurons. The evolution time will be 100 times as long with a single CAM8 box. With ten boxes, each with a separate microprocessor attached, to measure the fitness of the evolved network, the evolution time would be about eight minutes. Thus for 1000 neurons, the evolution would take about 3.5 hours, quite an acceptable figure. For a million neurons, the evolution time would be nearly five months. This is still a workable figure.

However, since the CAM-Brain research project will continue until the year 2001, we can anticipate an improvement in the speed and density of electronics over that period. Assuming a continuation of the historical doubling of electronic component density and speed every two years, then over the next eight years, there will be a sixteen-fold increase in speed and density. Thus the CAM-2001 box will be able to update at a rate of 200*16*16 million cells per second. To evolve the million neurons above will take roughly 13.6 hours. Hence to evolve a billion neurons, will take about nineteen months, again a workable figure. However, if a million neurons can be successfully evolved, it is likely that considerable interest will be focused upon the CAM-Brain approach, so that more and better machines will be devoted to the task, thus reducing the above nineteen month figure. For example with 100 machines, the figure would be about two months. The above estimates are summarized in Figure 10.4. These estimates raise some tantalizing questions. For example, if it is possible to evolve the connections between a billion artificial neurons in a CAM2001, then what would one want to do with such an artificial nervous system (or artificial brain)? Even evolving a thousand neurons raises the same question. One of the aims of the CAM-Brain research project is to build an artificial brain which can control 1000 behaviors of a robot kitten. Presumably it will not be practical to evolve all these behaviors at once. Most likely they will have to be evolved incrementally, i.e. starting off with a very basic behavioral repertoire and then adding (stepwise) new behaviors. In brain circuitry terms, this means that the new neural modules will have to connect up to already established neural circuits. In practice, one can imagine placing neural bodies (somas) external to the established nervous system and then evolving new axonal, dendral connections to it. The CAM-Brain project hopes to create a new tool to enable serious investigation of the new field of incremental evolution. This field is still rather virgin territory at the time of writing. The CAM-Brain project, if successful, should have a major impact on both the field of neural networks and the electronics industry. The traditional preoccupation of most research papers on neural networks is on analysis, but the complexities of CAM-Brain neural circuits, will make such analysis impractical. However, using genetic programming, one can at least build/evolve functional systems. The electronics industry will be given a new paradigm, i.e. evolving/growing circuits, rather than designing them. The long term impact of this idea should be significant, both conceptually and financially.

Sparc10	CAM8	CAM8	CAM8	CAM8	CAM2001	CAM2001
10000 CA cells	10000 CA cells	1 million CA cells	25 million CA cells	25 billion CA cells	25 billion CA cells	25 trillion CA cells
4 neurons	4 neurons	40 neurons	1000 neurons	1 million neurons	1 million neurons	1 billion neurons
1 Sparc10	1 CAM8	10 CAM8s	10 CAM8s	10 CAM8s	10 CAM2001s	100 CAM2001s
4.6 days	50 seconds	8 minutes	3.5 hours	5 months	13.6 hours	2 months

Figure 10.4 Evolution times for different machines, CA cell, neuron and machine numbers

10.6 FUTURE RESEARCH

A lot of work remains to be done. At the time of writing, the 2D version of a software simulation of CAM-Brain is nearing completion (having added nearly 7000 handcrafted CA state transition rules). Initial results are shown in Figures 10.5 and 10.6. Unfortunately, on a Sun Sparcstation 10, the time necessary to grow a sixteen neuron module is too long to be practical. Therefore it will be necessary to transfer the program to our CM5 supercomputer. Initial evolutionary experiments will be carried out on a four neuron module on the Sparc10 to see how well it evolves. As mentioned earlier, we already have experience with evolving simulated fully connected artificial neural modules (e.g. with sixteen neurons) and have found them to be highly evolvable [2]. Even if one cuts 70% of their evolved weights (i.e. one reduces their value to zero), the evolved function of the module remains more or less intact. Therefore, it is expected that CAM-Brain's neural circuits should still be evolvable despite less than full connectivity. If the 2D version shows good evolvability (i.e. if the CA based neural net modules evolve successfully to perform desired functions) then the next step will be to try a 3D version.

Since collisions occur more easily in 2D, a 3D version will be qualitatively different. To help visualize the cubes of the 3D cellular automata space in order to conceive the 3D CA state transition rules (using a six neighborhood transition rule, i.e. up, down, N, E, S, W), a Silicon Graphics machine will be used. If the 3D version also proves to be evolvable, then work will begin in earnest on the design of cellular automata machine hardware adapted to CAM-Brain. Contact has already been made with the CAM building group at MIT [7] to buy one of their recent CAM8 machines.

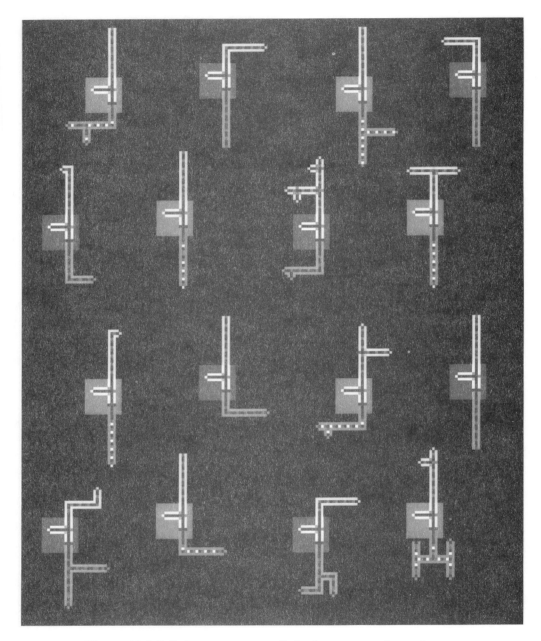

Figure 10.5 Cellular automata trails in the process of growing

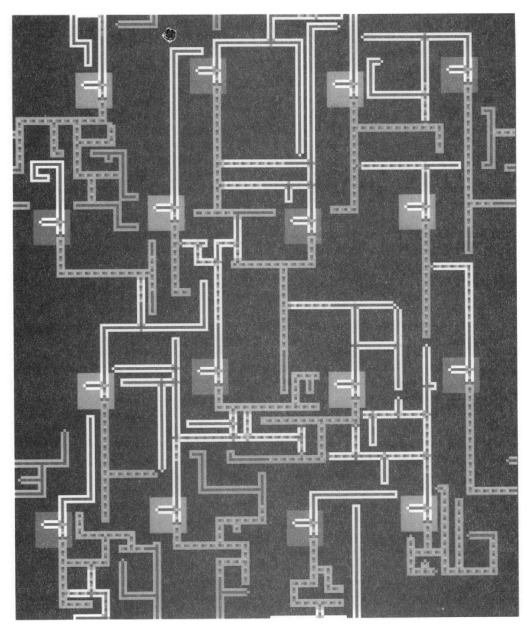

Figure 10.6 A grown CA network behaves as a neural network with measurable fitness

The beauty of using CAs as the basis for the CAM-Brain Project is that they allow the initial growth of a structure which can later be used. Thus, in effect, one has a type of evolvable hardware (EHW) [8]. There are two broad categories of evolvable hardware approaches, called intrinsic and extrinsic. In intrinsic EHW, the evolution occurs inside (intrinsic) the hardware itself, e.g. FPGAs (field programmable gate arrays, or other kinds of programmable logic devices (PLDs)) can have their circuit configuring bitstrings be conceived as chromosomes in a genetic algorithm, so that one obtains a new hardware circuit for each chromosome for each generation of the GA algorithm. In a CAM, the underlying hardware does not change, so strictly speaking, one is not doing EHW, but since a circuit gets evolved for each chromosome (where a chromosome is the sequence of signal cells which move down the middle of the CA trails), the process is equivalent to EHW. With extrinsic EHW, one uses software to simulate the evolution of a hardware circuit, e.g. by evolving a high level symbolic circuit description (e.g. using an HDL (hardware description language)), and then writing (down loading) the elite chromosome's solution into the configurable hardware. Hence the configurable circuit is written to just once. The real evolution occurs outside (extrinsic) the hardware. The CAM-Brain Project can thus be looked upon as a type of intrinsic EHW. Another feature of CAM-Brain, is that it will be possible to grow connections incrementally between one neurite module (i.e. a GenNet = genetically programmed neural network [1]) and another, and thus build/evolve/GP an artificial brain with thousands of GenNets. This would be a kind of incremental evolution. Since most papers on neural networks are concerned with only a single neural module, to be dealing with thousands of modules, as may be the case with the CAM-Brain Project, will be a breakthrough in neural network research. If successful, it will create a tool within which to build artificial brains. Initially, these artificial brains will contain only a small number of modules, e.g. 100s, later 1000s etc. Gradually, as the number of modules increases, and insight is obtained into how to evolve them incrementally, building artificial brains of increasing intelligence will become possible, thus creating a bridge between the two fields of artificial life (ALife) (whose long term goal is to create artificial life forms) and the more distant goal of artificial intelligence (AI) (whose long term goal is to create artificial intelligences). A sufficiently intelligent artificial life form becomes an artificial intelligence. Commercially speaking, if electronic circuits can be successfully grown/evolved, rather than be humanly designed, it may be possible to create circuits of enormous complexity and, hopefully, superior functionality. If so, the notion of evolvable hardware may revolutionize the electronics industry. By the year 2001, the last year of the CAM-Brain Project, our Brain Builder Group hopes to have the capacity to put a million neural net modules (GenNets) into a trillion cell cellular automata machine. During the 1990s, the Brain Builder Group also intends to design/simulate molecular scale CAMs and similar machines, so that the number of neural modules which can be evolved can jump from millions to Avogadro's number (i.e. a trillion trillion) once nanotechnology becomes a reality in the early 2000s [9].

10.7 CONCLUSIONS

This chapter introduced the CAM-Brain project and some of the implementation issues it is currently facing. More interestingly perhaps, it extrapolated future CAM-Brain processing abilities (to the year 2001 - the end of the current research project). Figure 10.4 shows that evolving artificial brains with millions, even billions, of artificial neurons within ten years is a serious possibility. This in itself is exciting. The real impact of CAM-Brain-like systems will be concerned with the issue of incremental evolution. CAM-Brain will provide a powerful tool with which to investigate such a question.

REFERENCES

[1] Hugo de Garis, 'Genetic programming: modular evolution for Darwin machines', ICNN-90-WASH-DC (*Int. Joint Conf. on Neural Networks*),   January 1990, Washington DC, USA (1990).

[2] Hugo de Garis, 'Genetic programming', in *Neural and Intelligent Systems Integration*, ed. Branko Soucek, Wiley (1991).

[3] Hugo de Garis, *Artificial Embryology: The Genetic Programming of an Artificial Embryo*, Wiley (1992).

[4] Hugo de Garis, *Genetic Programming: GenNets, Artificial Nervous Systems, Artificial Embryos*, Wiley (1994).

[5] Codd, *Cellular Automata*, Academic Press (1968).

[6] D. Goldberg, *Genetic Algorithms in Search, Optimization, and Machine Learning*, Addison-Wesley (1989).

[7] Toffoli and N. Margolus, *Cellular Automata Machines*, MIT Press (1987).

[8] Hugo de Garis, 'Evolvable Hardware : Genetic Programming of a Darwin Machine', in *Artificial Neural Nets and Genetic Algorithms*, R.F. Albrecht, C.R. Reeves, N.C.Steele (eds.), Springer-Verlag (1993).

[9] Drexler, *Nanosystems: Molecular Machinery, Manufacturing and Computation*, Wiley (1992).

11

Robustness of a distributed neural network controller for locomotion in a hexapod robot +

Hillel J. Chiel[1], Randall D. Beer[2], Roger D. Quinn[3], and Kenneth S. Espenschied[3]
[1] Department of Biology and Department of Neuroscience, Case Western Reserve University, Cleveland, OH 44106
[2] Department of Computer Engineering and Science and Department of Biology, Case Western Reserve University, Cleveland, OH 44106
[3] Department of Mechanical and Aerospace Engineering, Case Western Reserve University, Cleveland, OH 44106

ABSTRACT

A distributed neural-network controller for locomotion, based on insect neurobiology, has been used to control a hexapod robot. How robust is this controller? Disabling any single sensor, effector, or central component did not prevent the robot from walking. Furthermore, statically stable gaits could be established using either sensor input or central connections. Thus, a complex interplay between central neural elements and sensor inputs is responsible for the robustness of the controller and its ability to generate a continuous range of gaits. These results suggest that a biologically inspired neural-network controller may be a robust method for robotic control.

11.1 INTRODUCTION

In the last twenty-five years, much effort has been directed towards the construction and control of legged robots that might be able to negotiate complex terrains inaccessible to conventional wheeled vehicles [1]-[3]. A major obstacle that has slowed progress in this area has been the difficulty of controlling and co-ordinating a large number of multi-jointed

---

+   © 1992 IEEE. Reprinted with permission from *IEEE Transactions on Robotics and Automation*, Vol. 8, No. 3, June 1992, pp 293-303.

legs using centralized control approaches, where all decisions are based on all sensory inputs and all performance requirements. Recently, several investigators have attempted to utilize controllers which have been inspired, to a greater or lesser degree, by the neurobiological control of locomotion in insects [4]-[6]. The distributed nature of such controllers suggests that they might also be very robust to perturbations. Indeed, insects are capable of continuing to locomote despite wide variations in terrain, external loads, and damage to their bodies [7]-[9].

Robustness is especially important if legged robots are to function effectively in complex and potentially hostile environments. Robustness has two aspects. First, it implies that a robot can cope effectively with unexpected changes in its environment such as rough or uneven terrain. Second, it implies that a robot can also cope with damage to itself. Both aspects of robustness are extremely important, and ultimately both must be incorporated into robots that can function autonomously. In this chapter, we have focused on the second aspect of robustness, the ability of a robot to cope with damage. If an autonomous robot suffers the loss of a sensor or an effector, it should be able to continue to function. Similarly, if its controller is physically distributed across its body, it should continue to function if communication between local controllers is disrupted or a local controller fails.

Sensitivity analyses are often used to determine the robustness of controllers. Techniques such as robust-outer-loop control or adaptive control permit centralised controllers to cope with noisy sensor input, reduction in controller order as a consequence of the loss of an effector, parameter uncertainty, or changes in the properties of the environment (e.g., its stiffness) [10], [11]. Failure of components, which is equivalent to reducing the input of a sensor or the output of an effector to zero, is a particularly stringent test of the sensitivity of a controller to perturbation. Such failures are generally catastrophic for a centralized controller, while more distributed controllers should show a more graceful degradation of their performance in response to such failures.

In previous work, we have described a neural network controller for walking that is based on insect neurobiology and is highly distributed [6]. Companion papers describe a hexapod robot whose locomotion is directed by this controller [12, 14]. As a consequence of its distributed nature and the properties of its components, a steady input signal to the controller yields rhythmic patterns of leg movements. By increasing the intensity of the input signal, the controller generates a continuous range of patterns of leg movements that are not prespecified by the control signal, are statically stable and move the robot at increasing speeds, as shown in Figure 11.1. In Figure 11.1, leg labels indicate the side of the robot (L = left, R = right); numbers indicate position (1 = front, 2 = middle, 3 = rear). In this and all subsequent diagrams showing the robot's movements, the top six traces show leg angle (higher values correspond to forward position, lower values correspond to backward position). These traces show the outputs of the potentiometers on the robot, indicating the actual position of its legs. For each trace, the upper dotted line represents the angle at which the forward angle sensor goes on, and the lower dotted line represents the position at which the backward angle sensor goes on. The bottom six traces represent leg swing/stance phases schematically. For each leg, the dark bars represent the duration of the swing phase, while the space between the bars represents the duration of the stance phase.

These traces are generated by the neural network controller. The duration of data in this diagram and in Figures 11.5, 11.6, 11.7 and 11.8 is 8 s.

These results have suggested that a biologically inspired controller may provide significant advantages for the control of hexapod locomotion, in comparison to more centralized approaches that are more commonly utilized. Another consequence of the distributed nature of the controller should be an ability to cope with perturbations such as component failure. To test the robustness of this controller to component failure and to clarify its function, we subjected it and the robot to a variety of perturbations. Some of these perturbations were based on an earlier study of the robustness of a simulation of the controller embedded in a kinematic model of a hexapod [14]. The physics of the simulated world were greatly simplified, excluding the effects of friction, and treating force as proportional to velocity, rather than acceleration.

Since the robot has inertia, friction, delays in response to the controller, and other real-world properties that were not simulated, the controller might not be equally robust in an actual robot. Furthermore, the robot's mechanical structure can be perturbed in ways that were not possible in simulation. We will show that the robot and its controller are remarkably robust to perturbations as a consequence of the distributed design of the controller and the mechanical stability of the gaits that it generates.

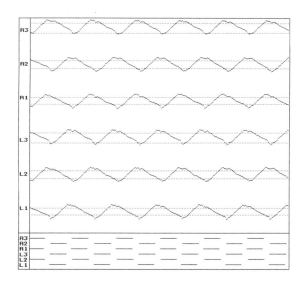

Figure 11.1 Normal gaits generated by the robot and the controller. (a) High-speed gait, also known as the tripod gait since each tripod of legs alternatively supports the weight of the robot while the other tripod of legs swings forward. Duration of leg cycle: 1.35 s.

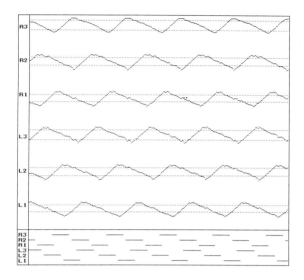

Figure 11.1 (contd). (b) Medium-speed gait; duration of leg cycle: 1.61 s.

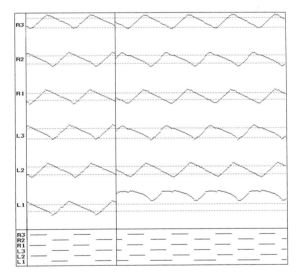

Figure 11.1 (contd). (c) Low-speed gait; duration of leg cycle: 2.18 s.

11.2 EXPERIMENTAL APPARATUS

The robot hardware and the controller are described in detail in [12]. The robot is 50 cm long by 30 cm wide and weighs approximately 1 kg (see Figure 11.2). Each of the robot's six legs has two degrees of freedom that are controlled by separate 2 watt dc motors (one to lift or lower a leg linearly, one to rotate it forward or backward in the vertical plane). Potentiometers attached to the motor shafts provide position information for both degrees of freedom. The neural network controller was simulated in C on a PC, and communicated with the robot via D/A boards; in turn, sensor inputs from the robot were sent to an A/D board and then read by the software. The controller generated vertical position and horizontal velocity signals. Velocity signals were integrated to yield a new leg position.

Figure 11.2 Photograph of the robot

Since the locomotion controller (described below) assumed that the swing and lift motions of the leg were independent, trigonometric transformations were applied to the controller outputs to generate appropriate radial and angular outputs of the motors. These outputs were fed into position controllers that drove the motors with voltages proportional to the deviations between the actual positions and the desired positions. The position controllers were implemented in analog hardware for speed. Because the motors were back-driveable, these controllers also gave a spring-like property to the legs of the robot, which was important for its ability to maintain posture.

The locomotion controller has been described previously [6,15]. Its architecture is based on studies of the neural control of insect-locomotion by Pearson and his colleagues [16]. The controller consists of thirty-seven model neurones: three motor neurones, two sensory neurones, one pacemaker neurone for each of the six legs, and one command neurone. The state of each model neurone is governed by the equation

$$C_i \frac{dV_i}{dt} = \frac{-V_i}{R_i} + \sum w_{ij} f_j(V_j) + INT_i + EXT_i$$

where V_i, R_i, and C_i represent the voltage, membrane resistance, and membrane capacitance of the ith neuron, w_{ij} is the strength of the connection from the jth to the ith neuron, f is an activation function that increases linearly above threshold until reaching saturation, EXT_i is the external current injected into the neuron, and INT_i is a time- and voltage-dependent intrinsic current that causes the six pacemaker neurones to oscillate. The period of a pacemaker's oscillation depends upon the level of synaptic input, with excitation decreasing and inhibition increasing this period. In addition, a brief inhibitory pulse occurring during a burst, or a brief excitatory pulse occurring between bursts, can reset the timing of a pacemaker's bursts.

A controller for a single leg is depicted in Figure 11.3. The forward swing motor neuron moves a leg forward, the backward swing motor neuron moves it backward, and the foot motor neuron raises or lowers the foot. When a leg moves forward, it activates the forward angle sensor neuron, and when it moves backward, it activates the backward angle sensor neuron.

During forward walking, legs alternate between a swing phase, during which the foot is raised and the leg swings forward, and a stance phase, during which the foot is down and pushes back. The command neuron provides steady excitation to the backward swing motor neurones, and (in the absence of other inputs) the foot motor neuron is active and keeps the foot down. Thus, the steady input signal from the command neuron will tend to keep the legs in the stance phase, steadily pushing backward. The stance phase is periodically interrupted by activity in the pacemaker neuron, which causes the leg to lift and swing forward by inhibiting the backward swing motor neuron and foot motor neuron while exciting the forward swing motor neuron. Three factors influence the duration of the pacemaker activity. First, when the leg moves backward, the backward angle sensor neuron is activated, which in turn excites the pacemaker neuron. Second, the pacemaker neuron has an important intrinsic property: once it is excited to burst, it does so for an essentially fixed period of time. Third, as a leg swings forward, it activates the forward angle sensor, which inhibits the pacemaker and thus helps to terminate its burst. The leg is then rapidly

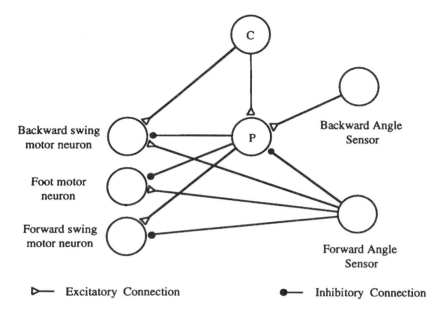

Backward swing
motor neuron

Foot motor
neuron

Forward swing
motor neuron

Backward Angle
Sensor

Forward Angle
Sensor

▷— Excitatory Connection   ●— Inhibitory Connection

Figure 11.3 Leg controller
(C: command neuron   P: pacemaker neuron)

switched from the swing to the stance phase by the activation of the forward angle sensor, which inhibits the forward swing motor neuron, excites the foot motor neuron (causing the foot to go down), and excites the backward swing motor neuron.

Increasing the activation of the command neuron affects the controller in two ways. First, it increases the intensity of activation of the backward swing motor neuron so that a stancing leg moves backward more rapidly. Second, it excites all six pacemakers, which increases their burst frequency as a consequence of the voltage-dependent current described above.

Co-ordination of the legs is achieved by inhibitory connections between adjacent pacemakers, which reduces the likelihood that adjacent legs will swing at the same time and compromise the robot's stability (Figure 11.4). Finally, slightly lowering the frequency of bursting of the rear pacemakers by increasing the angle ranges of the rear legs compared to the middle and front legs leads to reliable patterns of coupling between the pacemakers. Under these conditions, the rear pacemakers entrain the middle and front ones, generating a metachronal wave of stepping in which a rear leg swings before a middle one and a middle leg swings before a front one.

To study the robustness of the controller to damage, a variety of perturbations were applied to it. Since the controller and the robot are bilaterally symmetric, and all its local controllers are identical, a small set of perturbations could be used to predict the

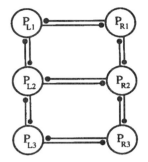

Figure 11.4 Pacemaker coordination
(Legs controlled by each pacemaker are indicated by subscripts)

robot's response to all possible single perturbations. We examined the response of the robot to five different kinds of damage: (1) failure of sensor input, (2) failure of effector output, (3) failure of communication between local controllers, (4) failure of local controllers, and (5) delays in response to commanded position changes. The first and third kinds of failure were studied by disabling connections from the sensors to the controller or by disabling connections between the pacemakers. The second kind of failure was studied by disabling the lift motor for a middle leg (its controller board was removed) and retracting the leg so that it could not support weight.

The fourth kind of failure was studied in several different ways. To study the effects of failure of a central local controller, a single pacemaker was completely disabled. To study the effects of failure of a peripheral local controller, the stiffness of the position controller was varied, as were the leg angles and their trajectories. Stiffnesses of the legs, which were set by the position controllers, were adjusted using potentiometers on the controller boards for the motors. Leg kinematics were varied by changing the appropriate program parameters. The fifth kind of failure was studied by altering the stiffness of different legs, which changed the speed with which they assumed a commanded position, and thus introduced delays to the controller.

We also studied the role of sensor input and central connections in gait generation by using several perturbations. First, we determined whether sensor input alone could generate gaits. To do this, we disabled the command neuron, and a fixed leg velocity value (set by keyboard input) was integrated by the program and used to update the position controllers for each leg. Second, we determined whether central connections alone could generate gaits by disabling all sensor input to the controller. Finally, we determined whether sensor inputs crucial for causing pacemakers to phase lock would differentially disrupt high- and low-speed gaits by disabling inputs for the backward angle sensors for both rear legs. Finally, we determined the role of the command neuron in stabilizing a gait by disabling its connections to the six pacemaker neurones.

11.3 RESULTS

11.3.1 Robustness of robot and controller

A robust controller should be capable of sustaining the loss of sensors, effectors, or central connections. The effect of such losses is illustrated in Figure 11.5. Removal of a forward angle sensor of any one leg led to exaggerated forward swings of that leg (see Figure 11.5(a)) but did not prevent the robot from walking stably at slow, medium, or fast gaits. In Figures 11.5(a) and 11.8(a), data to the left of the line show the pattern of movements before the sensor was disabled; data to the right of the line show the pattern after the sensor was disabled. The program allows a user instantaneously to pause the controller and the robot, save the screen that has been displayed up to that time, apply a perturbation, and then continue scrolling data on the remaining portion of the screen.

Since we wished to look at the steady-state response to the lesions, we would allow several screensful of data to be displayed, and then save the entire screen for printing. In this experiment, command neuron input was maximal, and the robot walked stably at the the highest speed gait (one can see alternate tripods of legs swinging or stancing in the bottom traces). After disabling the forward angle sensor for Ll, note that this leg moved much further forward and that the duration of its swing phase increased. Despite this change, the robot walked stably at this gait and all slower gaits.

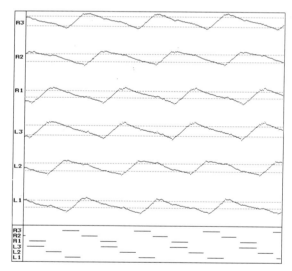

Figure 11.5. Effect of disabling single sensors, effectors, or central connections. (a) Forward angle sensor of left front leg (Ll) was disabled after the robot was walking at a high speed (tripod) gait.

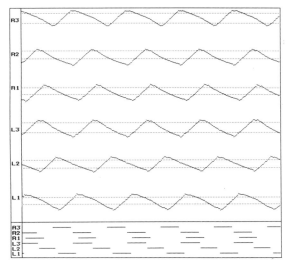

Figure 11.5. (contd). (b) Effect of disabling inhibitory connections from the pacemaker for foot R2 to the pacemaker for foot Rl.

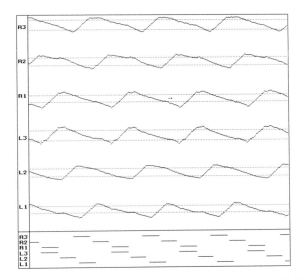

Figure 11.5. (contd). (c) Effect of disabling the lift motor of the middle left leg (L2) so that it could not support any load.

Similar results were observed after disabling the forward angle sensors for leg L2 or leg L3. Similarly, removal of a backward angle sensor of any one leg did not prevent the robot from walking stably at any speed, although gait patterns were significantly altered at slower speeds.

Removing any one of the central connections between pacemaker neurones also did not prevent the robot from walking stably at any speed, though removal of connections from a rear pacemaker to a middle pacemaker, or from a middle pacemaker to a front pacemaker caused the swings of the affected legs to overlap somewhat (see Figure 11.5(b)). Note that the swing phases for these legs now overlap (compare the swing phases for legs L2 and L1). Despite this change, the robot walked stably at this medium-speed gait, at slower gaits, and at the high-speed (tripod) gait. Similar effects on gait patterns were seen after disabling the inhibitory connection from the pacemaker controlling leg R3 to the pacemaker controlling leg R2. No effects on gait patterns or on the stability of walking were seen after disabling any other inhibitory connection between pacemakers.

Finally, disabling the lift motor of a middle leg and retracting the leg so that it no longer supported any load did not prevent the robot from walking stably at the slower gaits (see Figure 11.5(c)), although the robot was unable to walk at the fastest gait, which depends on one tripod of legs providing support while the other tripod of legs swings forward. If the leg was disabled but allowed to contact the ground, its forward movement tended to slow the movement of that side of the robot so that it turned toward the side with the disabled leg. Nevertheless, the robot was able to walk stably at low-, medium-, and high-speed gaits.

To explore the robustness of the robot further in response to the failure of discrete local controllers, we disabled single pacemaker neurones. The pacemaker of the middle left leg was disconnected from all other pacemakers, and the leg was disabled and retracted, as described above. The robot was not able to walk stably at the fastest (tripod) gait, but did walk stably at the medium- and slow-speed gaits. We did not examine the effects of disabling the forward or backward swing motor neurones, since removing these neurones caused a leg to swing to a mechanical extreme, which could have damaged the leg or its motors. By disabling the leg lift motor, as described above, we had determined the effect of disabling the foot motor neuron. In general, these investigations demonstrated that slower gaits are more mechanically robust than the tripod gait.

We also explored the effects of varying the local peripheral control (i.e., direct connections from sensor input to effector output) on the function of the robot. If the stiffness of the position controller was increased more than twofold, the leg position repeatedly overshot the position commanded by the controller, leading to oscillations. If the stiffness of the position controller was decreased more than twofold, significant delays occurred in the responses of the leg to the commanded position, which tended to spread out the gaits. The robot was able to walk forward and maintained its static stability despite these delays. A second source of reflex control was the direct connection from the forward angle sensor to the motor neurones for the foot and leg. Removing these connections did not affect the static stability of the robot at the tripod gait or at the medium-speed gait, but did lead it to become unstable at the slowest gait.

Since our position controller was based on a kinematic rather than a dynamic model of the robot, we did not explore the effects of applying mechanical loads to the robot or to its legs, or changing the irregularity or the slope of the terrain, which would significantly alter leg loading. We did explore two variations in the leg kinematics, which could have occurred if the position controller properties had altered. Varying the leg angles of different legs disrupted gaits slower than the tripod gait, since the stable coupling of the pacemakers depends upon these angle ranges. We also explored two different trajectories for foot contact during stance: one in which the foot moved in an arc and a second in which it remained level with the ground as it moved backward. In both cases, the robot walked stably, but it walked more smoothly (i.e., showed less side to side rocking) in the second case.

Finally, we observed that, in all gaits, the metachronal wave on one side of the robot occurred at about 180° out of phase with the metachronal wave on the other side of the robot (see Figure 11.1). By changing the stiffness of all the legs on a side, we introduced asymmetric delays into the position controllers, which disrupted the 180° phasing. In these circumstances, we observed that the feet of the robot slipped more often, presumably because the velocity changes became more irregular so that the foot was moving faster or slower than the rest of the robot when it contacted the ground. These results suggest that maintaining the 180° phase relationship between the metachronal waves on either side of the robot increases its efficiency and the smoothness of its walking.

11.3.2 Role of sensor and central input in gait generation

One possible reason for the robustness of the controller is that there are two mechanisms for generating the stance/swing cycle of each leg: first, the rhythmic outputs of the pacemaker neurones that periodically cause stancing legs to swing; second, the sensor input that tends either to induce a swinging leg to stance once it moves far enough forward, or to induce a stancing leg to swing once it has moved far enough backward. To determine the roles of these different sources of leg activation in generating stable gaits, we disabled the central command neurone connection to all six pacemakers and all backward swing motor neurones. Under these circumstances, the pacemakers do not burst spontaneously, and so the robot is unable to walk at all. However, if the robot is steadily pushed forward by using a constant velocity signal to update the position of the legs that are down, these legs move back until they activate the backward angle sensor; this in turn excites the pacemaker neurone, causing the leg to swing forward. Once it has swung far enough forward, the leg activates the forward angle sensor, which then inhibits the pacemaker, lowers the foot, and excites the backward swing motor neurone, preparing the leg for the next stance phase. As a consequence of sensor input, the leg cycles through a normal swing/stance sequence. Furthermore, the inhibitory connections between adjacent pacemakers   (Figure 11.4) prevent adjacent legs from swinging together. Thus, a full range of statically stable gaits is seen as the robot is pushed forward at different speeds (Figure 11.6). A further question which then arises is: what gaits can the central connections generate in the absence of sensor input?

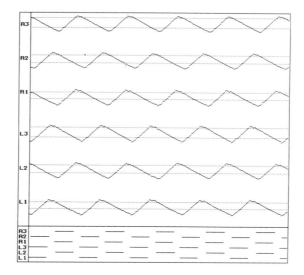

Figure 11.6. Gaits with leg activation using sensor input alone.   (a) High-speed gait.

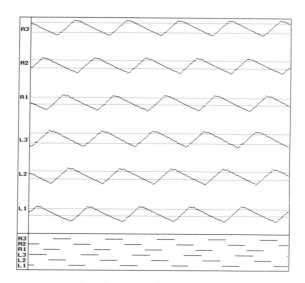

Figure 11.6 (contd). (b) Medium-speed gait.

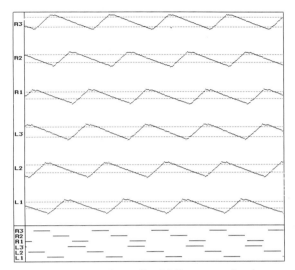

Figure 11.6 (contd). (c) Low-speed gait.

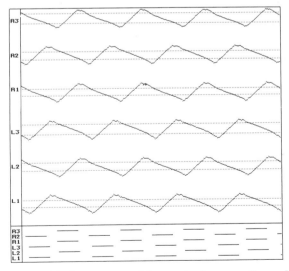

Figure 11.7 Gaits generated by central connections alone

When all backward and forward angle sensors were disabled, the robot was able to walk stably at a gait similar to the highest-speed gait (Figure 11.7). It was unable to walk at slower-speed gaits because its legs tended to move to their mechanical limits and were then unable to move further until inputs from other pacemakers changed the state of the pacemaker controlling them (i.e., turned it off or allowed it to go on by ceasing to inhibit it). These results suggest that the tripod gait may be generated primarily by central connections.

Less extreme perturbations of the controller demonstrated that its robustness was due to a complex interplay between sensor input and the central connections. An essential feature of all the gaits is that, on each side, a rear leg swings before a middle leg, and a middle leg swings before a front leg. Studies of the controller in simulation [14] suggested that this phase relationship between the legs is due to phase locking of the pacemakers.

As mentioned above, this phase locking is due to the slower rate of cycling of the rear pacemakers, which in turn is due to the larger angle ranges of the rear legs, which slows the rate of sensor input to the rear pacemakers. Thus, we predicted that removing sensor input from both backward angle sensors of the rear legs would severely disrupt slower gaits while leaving the tripod gait relatively unaffected. This was confirmed by the results shown in Figure 11.8. Despite the gait disruptions at the slower gait, the robot did walk stably. These results also suggest that many slower gaits that are nonmetachronal are also likely to be statically stable.

Another aspect of the subtle interplay between sensor input and central connections is provided by the transient behavior of the controller, as illustrated in Figure 11.9. Normally, within three to four swing/stance cycles for any given leg, the legs adopt a stable metachronal gait (Figure 11.9(a)). However, when connections from the command neurone to all six pacemakers are removed, leaving only its connections to the backward swing motor neurones, the controller requires six to eight swing/stance cycles for any given leg to settle into a stable metachronal gait and the transient behavior persists for much longer (Figure 11.9(b)). Even if the pacemakers are not stably locked into a particular gait configuration, the sensor input can bring them into a phase-locked configuration. Command neurone input brings them more rapidly to such a configuration but is not necessary if appropriate sensor input is available. Indeed, once stable gaits are established, pausing the simulation and removing all command neurone-to-pacemaker connections has no effect on gait patterns or the stability of the robot once it begins to walk again.

11.4 DISCUSSION

We have demonstrated that a biologically inspired controller for hexapod locomotion is remarkably robust. How does the controller work? Based on the results we report here and those obtained from an extensive perturbation study of the controller in simulation [14], we explain its function as follows.

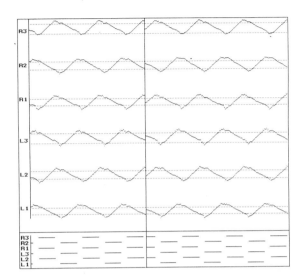

Figure 11.8. Effect of Partially removing sensor inputs   (a) Effect of removing backward angle sensors of both back legs on tripod gait

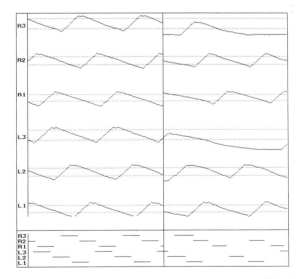

Figure 11.8 (contd)   (b) Low-speed gait pattern before (right of line) and after (left of line) sensor inputs from both backward angle sensors of both back legs were disabled

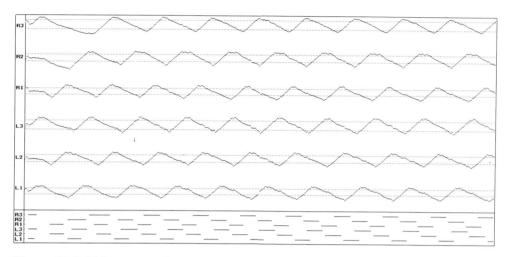

Figure 11.9 Using sensor input to compensate for the partial removal of the central command signal. (a) Transient response (time duration: 16 s) of the normal controller

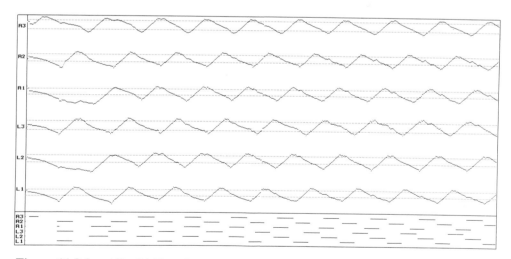

Figure 11.9 (contd). (b) Transient response (time duration: 16 s) of the controller when connections from the command neurone to all pacemakers have been disabled

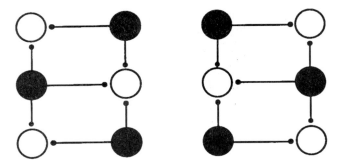

Figure 11.10 Generation of the tripod gait

The highest-speed gait, i.e., the tripod gait, is primarily due to the interactions between the six pacemakers. As one tripod of legs swings, the three active pacemakers controlling these legs inhibit the other three adjacent pacemakers, preventing the other tripod of legs from swinging. Once the swing has ended, the second tripod of legs can begin to swing, inhibiting the first set of pacemakers. Sensory input can reinforce this pattern but is not essential for its formation or maintenance. Figure 11.10 shows the two stable configurations of the central network during the tripod gait. In this diagram, neurons that are active are shown as black and inhibited neurones are shown as white.

In contrast, lower-speed gaits require sensor input both for their formation and for their maintenance. As gaits become slower, fewer adjacent pacemakers are active at any given time, and so maintaining the longer stance phase is due to two factors: the slower cycling of each pacemaker as the command neuron input to it decreases and the absence of excitatory input from the backward angle sensor due to the slower velocity with which the leg moves, since that input would switch the pacemaker on and terminate the stance phase. The coupling between pacemakers that generates the metachronal wave is due to the properties of the pacemakers and their response to inhibitory inputs.

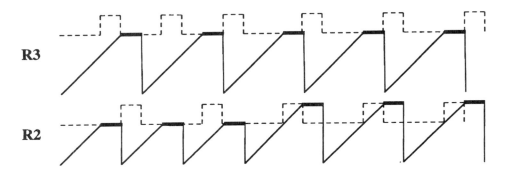

Figure 11.11 Generation of the metachronal wave

Generation of the metachronal wave is shown in Figure 11.11. In this, a heavy line represents the burst of a pacemaker and a dashed line represents the time at which a pacemaker's interburst interval terminates, which is increased when that pacemaker experiences inhibition due to the burst of the other pacemaker. Solid lines represent the progress of a pacemaker toward the end of its interburst interval. The slower pacemaker, R3, begins in phase with the faster pacemaker, R2. However, R3's inhibition of R2 acts to slow it sufficiently to entrain it after several cycles, resulting in a metachronal burst sequence. Thus, if one pacemaker is running at a slower rate than a second pacemaker, by inhibiting the second pacemaker it can delay when the second pacemaker turns on, thereby stably entraining the activity of the second pacemaker; the second pacemaker can in turn entrain the activity of a third, slightly faster, pacemaker. The gradient of angle ranges for the back, middle, and front legs results in a gradient of periods for the leg pacemakers, with the back legs having the largest period. As a consequence, the pacemakers controlling the back legs entrain the middle legs, which in turn entrain the front legs, resulting in the metachronal wave. Sensory input is essential for creating the gradient of pacemaker periods so that disrupting sensor input severely disrupts these slower gaits.

Despite the robustness of the controller, it has significant limitations. It is not designed to deal with changes in leg loading, although the position controller allows it to cope with small irregularities in the walking surface. Its ability to continue walking after a leg lift motor is disabled or after pacemaker neurones are disabled is not due to any intrinsic response of the controller but to the greater static stability of slower gaits, which allow more than three legs to support the robot's weight at any given time. As described in [12], it is also not designed to deal with asymmetric delays in the responses of legs to position commands. These delays lead to different lags in sensor input. The robot still walks stably and metachronal waves are observed on each side of the body, but the phase relationship between the two sides is abnormal so that one side is not 180° out of phase with the other side, as they are in the normal controller. Such delays are likely to occur in real robots whose mechanical and electrical components cannot be perfectly matched to one another. In addition, we have only dealt with straight-line locomotion on a flat terrain. Insects are known to be able to cope with extremely irregular surfaces [7] and to possess complex postural reflexes that allow them to maintain or regain their balance in response to mechanical perturbations of their legs or body [17]. In addition, insects can respond to leg amputations (which occur frequently in nature as a means of escaping predators) by reorganising their pattern of locomotion [8], [9]. It is even likely that agitated insects, or those who have had two legs amputated, no longer rely on a statically stable gait in order to locomote [9], [18], and thus their nervous systems may be capable of generating and controlling dynamically stable gaits, an issue that has been studied extensively by others [19]. We intend to utilise the growing literature on the neural mechanisms underlying these features of insect locomotion to guide the future design of locomotion controllers. Specifically, we intend to construct an insect-like robot whose legs each have at least three degrees of freedom. This will then be used to explore the role that postural reflexes in insects may play in allowing them to maintain their balance and traverse rough terrain.